Black 6

Black 6

The extraordinary restoration of a Messerschmitt Bf 109

Russ Snadden

PSL

Patrick Stephens Limited

Title spread: *Immediately following its painting, Black 6 looks like its old self, as found in November 1942 (see Chapter 16).*

First published in 1993

British Library Cataloguing-in-Publication Data:
A catalogue record for this book is available from the British Library.

ISBN 1-85260-425-5

Library of Congress catalog card no. 93-77246

Patrick Stephens Limited is a member of the
Haynes Publishing Group P.L.C., Sparkford,
Nr. Yeovil, Somerset, BA22 7JJ.

Typeset by BPCC Techset, Exeter
Printed in Great Britain by Hillman Printers (Frome) Ltd., Frome, Somerset.

Contents

Preface and Acknowledgements

Twenty years ago I volunteered to restore a little fighter plane to health. My friends were convinced that I was suffering from a form of temporary insanity — and, looking back, I would tend to agree with the diagnosis. Be that as it may, my intention was simply to return this historically important machine to a condition which would represent an accurate record of the type as it served in the Second World War. I fervently trust that in so doing I have illustrated that the ultimate success of my work has been largely due to patience, faith and the skill of my colleagues.

The self-imposed task took a shade longer than I anticipated! Along the way, however, I made many firm friends and I consider myself privileged to have led a unique team. Offhand I cannot think of any similar project involving an aeroplane of the class of the Messerschmitt Bf 109 in which all the personalities involved were unpaid volunteers. I can certainly think of no project of similar length!

In writing this book, I hope I have portrayed the many facets of the story — the moments of elation as well as those of despair, and the triumphs and the minor disasters. Without these the picture would be incomplete.

But it would also be incomplete without the endeavours of John Elcome, whose photographs adorn many of the pages. I am also indebted to John Dibbs, Patrick Bunce, Peter R. March, Mike Schoemann, Bobby Gibbes, Frank F. Smith, Keith Isaacs, Richard P. Lutz, Paulo Desalvo, Peter Patmore-West, Günter Behling, Heinz Langer, Don Batger, Richard Martin, Les Boot, Len Multimore, Frank Aldridge, Gavin Selwood, Brian Strickland, Ian Mason and my son Graeme for their photographic contributions.

My work is also embellished with several drawings — the work of John R. Beaman, altered, in part, by myself to suit the text. Some of the most accurate drawings of the Bf 109 yet published, he kindly permitted the use of a selection first seen in his book *The Last of the Eagles,* now out of print, sadly.

I must also thank the Public Record Office at Kew for allowing me to reproduce a report of test flying accomplished whilst the aircraft was in

Palestine, extracted from File Air 40/192.

I am indebted to Heinz Langer and the Lüdemann family for kindly allowing me to quote from the diary of Heinz Lüdemann and to reproduce pictures from his photo album.

In restoring this German aircraft, it was essential that I had a contact in its country of origin, and this role has been handsomely filled by Peter Nolte. Not only did he trace many items of equipment and arrange industrial assistance in his country, but he also translated a plethora of German manuals and documentation. Without this vital support, unstintingly provided, it is likely that the project would have spanned many more years.

Although his name features occasionally on the pages which follow, Andy Stewart deserves praise, not only for his physical help over the past ten years, but also for taking a keen interest in the history of the aeroplane. The research continues, but the career of the Bf 109 as presently known has been compiled with his enthusiastic help.

Since the debut of 'Black 6', my name has appeared in most of the resultant publicity, and I feel that the other personalities have been largely overlooked. It behoves me, therefore, to redress the balance and acknowledge their respective contributions. First, to Frank Shaw and Paddy Stanbury go my thanks for having taken an interest in a derelict heap, and for helping this pilot on an engineering road. For the years of sheer drudgery they spent in miserable circumstances, I am forever indebted to Kevin Thomas, Pete Hayward, Tony Leek and John Elcome. I doubt whether any others would have endured the hardship or displayed such remarkable patience.

At Bristol, where a remarkable and sustained effort produced a rejuvenated engine, my admiration goes to Ross Butler, George Ford, Roger Slade and Russell Stokes. However, without the sheer tenacity of John Rumbelow over the years, I suspect I would never have received the very necessary horsepower. He surely merits my deepest gratitude.

As the project drew towards completion, I consider myself fortunate to have found two such willing assistants in the shape of Bob Kitchener and Chris Starr.

The reconstruction of 'Black 6', though, is largely due to the skill of John Dixon and Paul Blackah. I have never ceased to admire their work, and I am under no illusion that, without their timely involvement, 'Black 6' would not have been completed.

Similarly, without the expertise, support and friendship of Ian Mason over more years than I really care to contemplate, the project would have foundered. In persevering when the task ahead seemed impossible to most, he sowed the seeds of a successful conclusion.

And, last, notwithstanding the dedication of all concerned, the restoration would have been impossible without the patient understanding of their ladies! For allowing the boys out to play, time and again, they each have my deepest gratitude and respect. To them, and particularly my own spouse, Phyl, this book is unhesitatingly dedicated.

1

First find an aeroplane

I took the decision to write this book many years ago. Now, as I put pen to paper, I regret that more positive action did not then ensue, for, with the passage of time, it is proving exasperatingly difficult to recall how I drifted into a sphere of interest which led, ultimately, to a certain fighter aircraft. I have been warned on many occasions that memory would be one of the (many) casualties of the ageing process, most recently by Jack Bruce, former Keeper of Aircraft at the Royal Air Force Museum and eminent researcher of the earliest days of powered flying machines. I hasten to add that, in counselling me on the fallibility of memory, he was not reflecting on a personal inability to remember these far off days! How remarkably easy my

Chivenor Open Day, and 10639 masquerades as 'Yellow 14'. A source of much amusement over subsequent years, it is astonishing how many Hispanos and 'Bf 108's have been painted to represent the mount of Hans-Joachim Marseille.

self-imposed task would be had I recourse to even the most sparse of diary entries.

My early interest in aviation was scarcely surprising. My father, an engineer with Rolls-Royce all his working life, bussed to the factory at Hillington. Close by lay Renfrew, the airport for Glasgow, which also functioned as a servicing base for Scottish Aviation, in those days heavily involved in the overhaul of Royal Canadian Air Force aircraft committed to NATO. I spent many happy hours admiring Canadair CL-13 Sabre and Avro CF-100 Canuck fighters on test. At home, some miles to the south, I was regularly reminded of the airfield when, each evening around seven, the synchronized sound of four superannuated Merlin engines passed overhead as a freight-laden Avro York clawed for height on its way south. A few miles east of Renfrew lay HMS Sanderling, alias Royal Naval Air Station Abbotsinch. On this airfield lay several hundred Fleet Air Arm aircraft withdrawn from service and awaiting their fates. Even today I can recall the sadness I felt climbing over the beautiful Seahawks, the Sea Venoms, Gannets and the old Avengers and Skyraiders, knowing that most would not survive more than a few months. It is probable that my eventual involvement in the embryonic preservation movement stemmed from this experience, and was, therefore, almost inevitable.

Abbotsinch lost its military role a few years later. Today it is Glasgow Airport. Of passing interest, perhaps, when I began flying into it after an enforced absence of some years, I noted a warning on the Jeppesen approach plates about the location of Renfrew Airport. Time and weather permitting, I pressed my nose to the windshield on many occasions in an effort to see it, to no avail. Subsequently, and at ground level, I discovered nothing remained of the old airport, and what used to be the main runway now forms part of the M8 motorway linking Glasgow with Edinburgh! I was not inclined to suggest to Jeppesen, though, that they remove this reminder of the past from their material.

Browsing through one of the aviation periodicals of the day, I chanced upon an invitation from a Malcolm Fisher to contact him with a view to sharing an interest in old aircraft. 'Bill' (as he is known, but the reason why escapes me!) proposed researching the histories of existing aircraft and, on joining him in this venture, I was made responsible for British military types. Over some months we discovered many old machines, some rare and others unique, which were in real danger of scrapping and, after much argument pro and con, we decided to try to save some of them. Thus evolved the ill-fated Historic Aircraft Preservation Society.

To appreciate the significance of this event it should be realized that, at the time, there was scant interest in the subject. In truth, examples of a selection of aircraft were scattered throughout the United Kingdom, belonging in the main to the Ministry of Defence and receiving minimal maintenance. (To this topic I will later return!) There was no Royal Air Force Museum and no Imperial War Museum, Duxford. Preceding HAPS, the Northern Aircraft Preservation Society had pioneered enthusiast involvement in the conservation of several notable aircraft. Its influence, though, was necessarily localized. HAPS aimed to be a national movement. We had no clear plans at the outset, merely the aim to save as many deserving machines as we could.

The first to 'come our way' was a Chance-Vought FG-1 Corsair Mk 4, serial number KD431. Employed for many years as an instructional airframe by the Cranfield College of Technology, it had been deemed surplus to requirements. An approach by HAPS bore fruit and we became the proud owners of this classic fighter — at the time the only one in Britain. It can be seen today in the Fleet Air Arm Museum, Yeovilton. (Several now fly on the display circuit.) Shortly thereafter, we discovered the last remaining Vickers-Supermarine Seafire FRMk47, VP441 in the less-than-tender care of an Air Training Corps at Plymouth. Then an eyesore after years of neglect and abuse, it was donated to HAPS provided its removal could be speedily organized. This we achieved with the help of the Navy. The aircraft was offered on loan to Culdrose Naval Air Station and transported there, receiving a badly needed refurbishment and a time-expired propeller assembly from an Avro Shackleton, the original having long since vanished.

What became of this aircraft is a cause of some concern. Sold some years later, it was shipped to the United States to be rebuilt (it was rumoured) with a view to attempting to break the long-standing speed record for piston-engined aircraft. Nothing has been heard since and it is most doubtful whether even this airframe/engine combination, representing as it does the last of the Spitfire family, could hope to better the present record. I still cherish the hope that it may return home — some day.

Aircraft followed aircraft, but only one was ever bought. I paid the handsome figure of £25 for the remains of Vickers-Supermarine Walrus Mk 1 amphibian, L2301! Located by the Society in tall grass on the perimeter of Thame Airfield, Oxfordshire, these consisted of a fuselage (minus all tail surfaces), engine and propeller unit. The Navy came to the rescue once again and, on loan terms similar to those on the Seafire, carried the hulk north to Arbroath in Scotland. The old bird was beautifully restored there, incorporating new-build tail and mainplanes, and today occupies a special corner of the FAA Museum, not far from 'our' Corsair.

Another large step into the unknown was taken when 'Bill' Fisher applied to the French Navy for one of a small number of Avro Lancaster Mk VII bombers being withdrawn from service in New Guinea. Much to our astonishment, his request was granted! To Bill's eternal credit, he organized help and sponsorship (mainly from Hawker de Havilland, Sydney and Shell Oil) which enabled NX611 to fly home, arriving at Biggin Hill during the Air Fair of 1965. Regrettably, the Lanc. proved too much for the meagre resources of HAPS. The stringent requirements of the Civil Aviation Authority to keep the old aircraft flying proved insuperable. Ultimately, therefore, it caused the demise of the Society, and the voluntary structure collapsed amidst much acrimony with the emergence of an overly-greedy portion of the membership. The aircraft survived, however. After some years adorning the gate of RAF Scampton, it is now owned by Fred and Harold Panton and, beautifully restored, it now forms the major exhibit of the Lincolnshire Aviation Heritage Centre at East Kirkby, Lincolnshire.

As an experiment, there is no doubt that HAPS was a failure. However, most of the aircraft it rescued have survived. I am convinced that most would have perished otherwise. For that reason alone I am proud to have been involved.

The world of old aircraft has changed since those exciting days. Many societies were subsequently formed and several survive to continue their good work. Overshadowing this volunteer effort, there emerged the new-born interest of moneyed individuals who espied investment potential in antique machines. Today, as a direct result, we have a plethora of aircraft maintained in airworthy condition and displayed regularly to the delight of the air-minded. However, the negative side to this situation is that, as a result of all too frequent sales of these machines (evincing, perhaps, a loss of interest by the owners?), values have been artificially inflated to the extent that only a few favoured people can afford them. The potential active involvement of so many people in *preserving* old aircraft has been sacrificed. No longer are redundant machines donated into the care of the enthusiast.

Despite the collapse of HAPS, my interest in old aircraft suffered only temporarily. I resolved, though, that any future project would be under my control solely. There would be no place for decision by committee. By then, I was flying De Havilland Comet CMk4Cs with Two-Sixteen Squadron, Royal Air Force Lyneham in Wiltshire, and operating in the VIP transport role. The task took me all over the World during a five year posting, and, in the main, flights were lengthy, sometime entailing being away from base for three to four weeks. There were similar periods between tasks and I reasoned that this lifestyle would permit me to undertake the restoration of a suitable, small aircraft. Selecting the lucky subject proved a tortuous experience. Whilst my RAF salary and flying pay permitted a not-uncomfortable life within the Service, it was far short of affluence! I might just have been able to buy an old wreck, but I could not have afforded the maintenance cost. The realization that I would be unlikely to actually own an aeroplane did not deter me. Today, my attitude would be somewhat different, I fear!

I have observed that most people pursuing a hobby with any zeal eventually develop an inexplicable affection for a small portion of their chosen interest. The car fanatic will favour a marque, or even a model, for example; a fisherman may gravitate towards casting a fly. So it was with me. My study of aviation history had focused on the military and, in particular, on the Second World War. Within that wide subject, I had become engrossed in the German Air Force and its equipment. I assume this was because so little remained of that once powerful force. But relics *did* exist from that era, most belonging to state organizations. In Britain we had a diverse collection of German aircraft which rivalled the inventory of the Smithsonian Institute in the USA. The die was cast; I decided that one of these rare machines would be the recipient of my enthusiastic attention. Much to my surprise, however, the authorities were far from convinced.

My first approach was to the Imperial War Museum. Within its grey walls in Lambeth lay a fine example of the magnificent Focke-Wulf Fw 190 fighter. On loan to IWM from the Ministry of Defence (a situation which exists to this day), the aeroplane had received little attention since its incarceration. (The camouflage scheme has recently been altered, but the new appearance is appallingly inaccurate — a great pity as, with only a modicum of knowledge and suitable research, a handsome restoration could have resulted.) To the suggestion that I be allowed to restore it, the IWM Director, Nobel Frankland, sent an instant rebuff. Disappointedly, and in no little frustration, I next tackled the real 'owners'. They disclaimed any influence or responsibility and referred me back to IWM! It was a train of events which was to prove familiar over the following months. The only other 190, a training conversion of the fighter, was in the care of the Royal Air Force at St Athan, South Wales. By this time (1970) The Royal Air Force Museum was taking shape. Aircraft in the theoretical charge of the Air Historical Branch, but in reality in the reluctant care of countless RAF stations, had been reserved for inclusion in the new building, or had been classified as parts of the 'Reserve Collection of the RAF Museum'. At St Athan, the staff had performed near miracles in preparing externally a selection of German types, all of which can be seen today at Hendon. The supreme example of their work is undoubtedly the Messerschmitt Bf 109 E-4 whose present appearance is the result of copious research by Squadron Leader Schofield (now retired). I was not surprised that my offer to restore the 190 was rejected. St Athan was losing a large percentage of its 'treasure' and was, I suspect, loath to part with more. But, more important, the training version of this classic fighter was produced in limited number. The sole surviving example could not be risked in a venture which would have seen it rebuilt to flight-worthy condition. It can be viewed at Hendon today, regrettably also clad in improbable colours.

Approach No.3 was aimed nearer to home. It was also less ambitious in that it targeted an aeroplane which could not be considered for rebuild to flying condition. A few miles west of Lyneham lay RAF Colerne, which, in the early '70s was home to a 'regional collection' of museum aircraft. A small band of airmen had just completed a voluntary restoration of the petite Heinkel He 162 A jet fighter. The result of their labours can be seen today at Hendon, although the formerly beautiful paintwork now bears many scars as a result of its dismantling and transportation. The cockpit is devoid of many components, unfortunately, but it is without doubt the most accurately restored example of this diminutive fighter, and an example of what can be achieved as a result of suitable research and the attention of the enthusiast.

Within the same collection was an unrestored example of the radical Messerschmitt Me 163 B interceptor. The only pure rocket-powered fighter ever to see service, it could climb to 39,000 feet, from a standing start, in 3½ minutes, after which, fuel exhausted, it was intended to wreak havoc on Allied bomber formations by gliding through them in attack, initially at speeds in excess of 500 mph, until height and speed decayed, forcing a rapid return to earth. Named Komet by its creators, I can conceive no better appellation to describe the fantastic sight of its powered climb. As I was flying the Comet airliner, I was confident that the connection would elicit support from my unit. The Station Commander readily approved my plan, but, in the midst of arranging transport for the little beast, a message arrived which scuppered the whole scheme. He refused to allow the aircraft to leave his domain!

W. Nr. 10639 arrives at Lyneham, its new home.

If my efforts thus far have conveyed the impression of 'beating my head against a brick wall' of officialdom, it was merely a foretaste of the main course. On reflection, I find my perseverance astonishing. It can be attributed only to unbridled enthusiasm — and perhaps a thick skin. I cannot deny, however, that my aspirations had been dealt several severe blows, but I was not yet prepared to accept defeat. It occurred to me that all three aircraft in which I had expressed an interest belonged to collections. The respective caretakers were, understandably, unwilling to part with them. One German aircraft, though, was held in isolation at RAF Wattisham, Suffolk. This was a Messerschmitt Bf 109 G fighter which had been held for some years following an abortive rebuild attempt.

It was apparent to me that negotiating for this aeroplane might prove complicated as three fingers were in this particular 'pie'. The Air Historical Branch of the MoD retained responsibility, but being largely toothless, it was totally dependent on the Royal Air Force to care for its museum pieces. In this instance, Wattisham would have to be wooed. I also anticipated the active interest of the RAF Museum. I submitted my plan to Group Captain Haslam, head of the AHB, early in 1971. His reply graciously thanked me for my interest but indicated, much as expected, that the 109 was held on behalf of the Museum. A second letter winged its way to Jack Bruce, Keeper of Aircraft and Aviation Exhibits at Hendon. Jack was largely sympathetic to my scheme, but felt that he could not release the aircraft into my charge as the Museum might plan to include it in future displays. Subsequent letters attempted to cajole and persuade both parties, but to no avail.

At that point, I confess, my resolve almost collapsed! After all, why should I demean myself by *begging* to work on their aircraft when it was all too obvious that they would rather I left them as they were. The year 1971 had been consigned to history before my batteries had recharged sufficiently for another assault. I wondered whether, in the intervening months, the plans of the

A nice view of 'Yellow 14', alias 10639, in sorry condition.

Museum had formalized. Progress with my plans hinged upon attitudes at Hendon; approval from that source would guarantee the success of my bid. A letter, little changed in content from one sent a year previously, found its way to the overburdened desk of Jack Bruce. This time, a small ray of hope emerged. Space was at a premium at Hendon, and there was certainly no room for more than one Bf 109. It was expected that the St Athan Bf 109 E-4 would be exhibited, paired with a Spitfire Mk 1. Jack indicated that he could not be seen to be helping me 'as it may be misconstrued by some'. However, he continued: 'I should love to see the 109 back in the air again, but the brutal fact is that circumstances are completely against my active participation in such a project at this time. I honestly feel, at this stage, all I can do is not to oppose your most worthy object and activities.'

These same words I transmitted to AHB, which reiterated that policy remained that the aircraft was 'earmarked for the RAF Museum'. Furthermore, it was revealed that Wattisham, somewhat belatedly, had plans for its 'maintenance', and intended displaying it at their Open Day that September. The Station Commander, Group Captain Goodwin, indicated that he *might* be prepared to part with it, but *only* after September! Months later he admitted to me that he had been saddened to see the deterioration that had overtaken the aircraft and that he thought it would be in everyone's best interests that I took it under my wing.

On Saturday 30 September 1972, two Hercules transport aircraft returned to Wattisham after an exercise in Denmark. To the surprise of the crews, rather than going home empty, they found a dismantled Messerschmitt awaiting them. Thus, that afternoon, the mortal remains of the fighter were unceremoniously disgorged on the apron at Lyneham. My telephone rang, and a voice demanded: 'Excuse me, Sir, but do you know anything about a Messerschmitt?' I had become the custodian of Messerschmitt Bf 109 G-2, Works Number 10639.

The Eagle from Messerschmitt

The Bf 109 has been the subject of books and articles beyond number. In this respect it is rivalled only by its British counterpart, the Supermarine Spitfire, and it is, perhaps, an indicator of the historical importance of the two fighters. The Spitfire, though built in great quantity, did not match the total production figure of the Messerschmitt which, we are told, exceeded 30,000 machines. Both represented astounding advances in fighter design, and generated great public excitement upon their appearance. Well over half a century later, this excitement is seemingly unabated!

Most important, they presaged the end of the biplane era, the Spitfire displacing the ultimate British design — the Gloster Gladiator — whilst the Bf 109 succeeded the sleek little Heinkel He 51. Gone, together with one wing, was the plethora of struts and bracing wires and, with only a modest initial increase in engine power, attainable speeds increased by almost 50%. Yet, remarkably, size changed hardly at all. Of low-wing, cantilever (externally unbraced) configuration, the Bf 109 was novel in featuring a retractable undercarriage and enclosed cockpit combined with all-metal monocoque construction.

In the biplane, the wings (or, more correctly, mainplanes) generally featured all-wood construction, the spars commonly fashioned from spruce to which were fastened wooden ribs, and the whole enclosed by fabric, doped to provide taughtness. Fuselage structure was composed of a rigid cage of metal tubes over which were placed wooden formers to give shape to the skin of fabric which was laid over, stitched and doped. The strength of the assembly, then, lay in this internal cage.

The monocoque reversed the situation, and strength was provided by the outer, aluminium alloy skin. In fact, the Bf 109 employed a semi-monocoque fuselage, some of the overall strength being supplied by an internal array of fore and aft frames, or longitudinal stringers. Behind the cockpit, the rear fuselage comprised two halves, left and right, each built separately. The first step of the process saw sheets of alloy rolled into shape. Alternate panels then had their front and rear edges formed into Z-section, these becoming integral framing and providing a deal of rigidity. These same edges were also worked into a step sufficiently shallow to affix adjoining unlipped panels and form a flush link, the whole being put together using counter-sunk riveting. Fuselage halves, bolstered with long stringers, were then mated, top and bottom. As far as I am aware, this method of construction was unique. It was remarkably simple, easy and quick to produce — an important factor in wartime.

The tail surfaces were just as quickly manufactured. Made also in two parts, the fin and tailplane halves were joined using a hinge pin in the leading edges and fastened together at the trailing edges — clam shell fashion. In recognition of the control inadequacy of the vertical tail and rudder in the slow speed realm, the fin was formed asymmetrically, providing an aero-dynamic force to deflect the tail to the left, partly counteracting the torque effect of the large propeller, particularly during the take-off run.

The forward fuselage contains the *'führerraum'* — the cockpit — and the fuel tank bay. Behind and beneath the pilot's seat is inserted a solitary, rub-berized, self-sealing fuel cell of 400 litre capacity, and forward of that lies the centre portion of the wing spar. On the forward lower corners, two complex steel frames were bolted. These became known to us as the 'D-frames', in due deference to their shape, and they are extremely important structures in that they mount the main undercarriage legs and house their vital down locks, provide a lug for a locating pin on the leading edge of each mainplane, and also support the lower engine bearers. A failure in either of these frames guaranteed severe damage to the aircraft. Mounting the undercarriage on the fuselage undoubtedly facilitated the movement and transportation of the fuselage. It also allowed removal of both mainplanes without recourse to trestling. The obvious drawback was that it resulted in a narrow track — but, having said that, it was no narrower than the Spitfire on which the legs were wing-mounted.

The mainplanes are beautifully simple in construction. The main spar is a solid alloy beam and is augmented by an auxiliary spar in the shape of the rear edge of the wing proper. Riveted to both are pressed alloy ribs, and the skin is joined to spar caps, again using countersunk riveting throughout. Automatic leading edge slats adorn the outer reaches of the leading edges. These devices were pirated examples of a Handley-Page invention intended to enhance lift characteristics of the wing, particularly at slow airspeeds, but also in manoeuvring. They open and close purely as a result of sensed airflow; the pilot has no control over them. Wing tips, originally square-cut fairings, had developed into detachable, elliptical assemblies by the time the later, more powerful versions arrived on the scene. Underneath, and com-mon to all Bf 109s, lay two coolant radiators, close by the wing roots and within the cooling arc of the propeller. These, too, had been progressively developed, and in the G-series aircraft featured an ingenious air intake and exhaust arrangement. At the rear of each fairing, and in line with the plain flaps positioned immediately outboard, are an upper and lower 'door'. Both of these radiator flaps are interconnected which, on opening, transmit that movement to a leading edge lip which opens in sympathy. Not only that, but the whole arrangement is linked to the plain flap! As the pilot lowers flap, the radiator flaps also depress but maintain the distance which separates them. Thus, flap area is increased, improving slow speed handling, whilst airflow through the radiator remains unimpeded to maintain coolant temperatures. In order to ease the workload of the pilot, temperature is thermostatically controlled. If required, he may dispense with this automatic function and, by use of a handle by his right leg directly linked to a rotary valve, 'inch' the flaps open or closed. However, hydraulic power is necessary for any control of the system. The Messerschmitt design team added a further refinement. Radiators were vulnerable to damage in combat and, if holed, quickly dumped all cooling fluid overboard, resulting in engine seizure and probable loss of the machine. It was realized that the influence of one radiator would be sufficient to allow a safe landing without engine damage. Thus, an 'O-handle' was fitted to each side of the cockpit which, when pulled, triggered shut two valves in the pipes supplying the stricken radiator. These did not eliminate the escape of coolant but, rather, reduced the rate of loss.

All flying control surfaces were of alloy construction but fabric-covered, and were to remain thus throughout the War: a strange anachronism for a company which was in the forefront of fighter development. Just as curious, pilot adjustable trim was provided only in pitch by altering the incidence of the tailplane. No controllable trim tab was ever fitted to the 109 aileron, whilst rudder trim was only introduced on much later versions which featured an extensively redesigned fin and rudder.

Power for the Bf 109 in its 'G' form is provided by the magnificent Daimler-Benz DB 605 engine driving a VDM (*Vereignte Deutsche Metallwerke*) forged alloy propeller. The principle of operation of this differs considerably from all contemporary designs. British and American constant-speed propellers in particular were controlled via a lever, additional to the engine throttle. The pilot selected an appropriate rpm, and a governor (known as a constant speed unit — CSU) maintained this rotational speed within its operating parameters by adjusting blade pitch, regardless of the power demanded by the pilot through use of the throttle. The VDM propeller dispensed with the rpm lever. Power from the engine is demanded by throttle position in the normal fashion, whereupon optimum rpm is automatically selected. Propeller rpm is then maintained, provided the throttle is not further disturbed. Thus, although a constant speed propeller, the principle of its operation is dramatically different from its contemporaries.

The Bf 109 provides its pilot with the option of further, more positive, control of his propeller. By use of a switch mounted below the throttle quadrant, automatic governing may be by-passed. In this instance, he can alter the pitch of the propeller blades directly by use of a rocker switch on the throttle grip, the new setting being selected by reference to a clock-like gauge on the instrument panel. The unit becomes multi/variable pitch, therefore, and reverts to a mode of operation best associated with earlier, more primitive designs.

The armament of the Bf 109 G in its early form was very lightweight and tended to be inferior to its opponents. It sported two 7.9 millimetre MG 17 Rheinmetall-Borsig machine guns mounted ahead of the windshield and synchronized to fire through the propeller arc. Bolted to the rear of the engine casing, however, its breach positioned between the rudder pedals in the cockpit, lay a Mauser MG 151 cannon of 20 millimetre calibre. This extremely effective weapon fired its ordnance through the tip of the propeller spinner. (The ideal position of this weapon, mounted precisely on the thrust line of the aircraft, was a triumph for the VDM team, as they designed the propeller and its gearing for that purpose. It was an ingenious solution, and one much admired in Britain.) Additionally, a further MG 151 could be attached below each wing, but in the early 'G' this was rarely fitted. By comparison, the Spitfire of the day (the Mk V), was armed with eight 0.303 inch Browning machine guns (or four such guns plus two 20mm cannon) but all wing-mounted and requiring careful synchronization to ensure the convergence of ordnance at the correct distance ahead of the aircraft.

A great deal of research by acknowledged experts suggests that the early Bf 109 G was slower by 20 miles per hour than the contemporary Spitfire. Designed as a high altitude fighter, the 109 was superior aloft by a similar margin. Whereas the 109 could easily out-climb the Spitfire, the British product could out-turn the German. Throughout the technical development of both machines, the performance of one leap-frogged the other, but it is patently obvious that, whilst even a tiny superiority in performance was critical to a pilot in combat, both aircraft were evenly matched. To my mind, the most interesting comparisons lie in the airframe. I have described the remarkably simple construction of the Bf 109 mainplane, for example. Aesthetically, there can be no argument that the mainplane of the Spitfire is beautiful and, aerodynamically, it was very efficient. The planform of Reginald Mitchell's elliptical masterpiece shows hardly a straight line anywhere. The graceful curves, though, complicated the production process. Whereas 109 ribs were pressed sheets of alloy, each Spitfire rib was constructed of a lattice of small strips riveted together. Whereas countersunk rivets were used throughout the German aircraft, great use was made of snap-headed rivets on the British. Easier to form, the proud, dome-shaped heads of these cover the rear fuselage and tend to mar the lines of the design. Messerschmitt and Mitchell each designed a fighter revolutionary in concept. Of similar configuration, they were very different in construction, but there is little doubt the Bf 109 was better conceived for ease of production.

Introduced into service by the Luftwaffe in 1937, the Bf 109 remained a front-line fighter until the end of the Third Reich which was responsible for its birth. Trial by combat was provided when a few early production machines supplanted obsolescent Heinkel He 51 fighters in supporting the Nationalist cause in the Spanish Civil War. They outclassed the mainly Soviet-built opposition. Equipped with a weapon of proven superiority, the Germans rapidly subdued opposing air forces. Although France had begun to produce the Dewoitine D.520, a modern fighter which might have caused the German Air Force some headaches, few reached service status as a result of chaotic production lines.

Across the Channel, though, sufficient numbers of Hawker Hurricanes and Spitfires had reached the Squadrons to give the Luftwaffe and its Bf 109s their first bloody nose. During and after the Battle of Britain, both the

Spitfire and Bf 109 were developed, engine power being increased and the respective airframes benefiting from aerodynamic improvements.

Second generation monocoque fighters helped tilt the balance of success towards the Allies, particularly with the arrival of the North-American P-51 Mustang and the Republic P-47 Thunderbolt. The Mustang was a superb piece of engineering, slightly larger and heavier than its European predecessors, whilst the Thunderbolt was an enormous machine hauled along by a massive radial engine of some 2,000 horsepower.

While American industry in particular had the foresight to develop these new fighters, the German High Command ordered into production only one new type. This was the Focke-Wulf Fw 190, whose appearance over the skies of France in 1941 both confused and shocked the Royal Air Force. The Fw 190 was powered by a BMW 801 radial engine, neatly cowled. In European terms it was unusual in concept as normal practice was to make use of the slimmer, less drag-inducing, water-cooled powerplant, but it outperformed all existing types by a handsome margin and rapidly established itself as a reliable and adaptable workhorse. The radial engine, however, rendered it most effective at low altitudes, and the Bf 109 remained the sole single-engined fighter of the Luftwaffe capable of tackling higher flying foes, at least until the late stages of the conflict.

The Messerschmitt company developed the 109 until the end of the War, but it became progressively more cumbersome, particularly with the addition of heavier armament, both internally and underwing, and its performance suffered as a result. The Luftwaffe pilots realized that new equipment was urgently needed, but the attitude of the *Reichsluftministerium* was somewhat different. In weighing the projected requirements of the Luftwaffe, it decided that the Bf 109, in improved form, would continue to be effective. The blinkered attitude seemed to be that, as it had performed well during the early years, it would continue so to do, and there was no point in spending time and resources developing a replacement. It was a short-sighted policy, and German pilots were fated to fly the same aircraft at the War's end, albeit in more potent form. Inferior though it was then, it was still a much-respected foe. Its service record was awesome and it remained the favoured mount of many pilots, despite displaying the obvious limitations of its elderly design.

3

The toil begins

My plan was to restore the fighter. 'Restore', however, is a word often misused, especially in the world of old aircraft. In 1973, 'completely restored' was a euphemistic description which had been applied to many projects. The reality was that the aircraft had benefited solely from yet another, and generally inaccurate, coat of paint. I had been less than impressed by these token efforts. My idea of restoration adhered more closely to the dictionary definition.

First, I determined that during the stripdown I would record any paint and markings found, so that they could be reproduced. Moreover, on the assumption that a substantial amount had been destroyed, I would under-

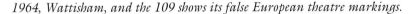
1964, Wattisham, and the 109 shows its false European theatre markings.

Our aeroplane immediately following application of spurious colour scheme at Wattisham in 1963.

take research to uncover the history of the machine and, perhaps, discover a photograph or two to aid the restoration. Second, I have always believed that restoration means the retention of as much of the original structure and equipment as possible. Otherwise restoration becomes reproduction, and that was a path I was not prepared to follow with the 109.

Another facet of my plans — a secondary aim, as it were — was that I would attempt to return the aircraft to an airworthy state. Should I have found any serious fault, though, that attempt would have ceased immediately as I was not prepared to modify the structure and, hence, compromise the restoration, merely to put it in the air. It may be argued that the latter objective should have been given priority, but it was very important to me that the old bird was restored. Also, the work was to be performed on a part-time basis and would receive no funding whatsoever from the Ministry of Defence. Being realistic, I was confident that I could restore the aircraft. I was less convinced that I could make it airworthy.

Viewing it for the very first time did not inspire me with confidence. On the contrary, that first sight appalled me, and the magnitude of my task started to become apparent. Despite growing apprehension, I managed to disguise the reality from my wife standing by my side. I felt that I appreciated what had to be done; I was sure she did not, which was just as well! The fuselage lay in front of us atop two main undercarriage legs of different lengths, with battered engine cowlings, detached from their mounts and draped over the engine, and a very bent rudder. Each mainplane had been damaged in similar fashion — tears and dents in the skins, missing navigation lights and torn aileron fabric. More serious, I noticed that the complex panelling which faired the wings to the fuselage was entirely missing. In their stead were sheets of cardboard!

It took several minutes to summon the courage to peer into the cockpit which bore an inappropriate *Erla Haube*. (This canopy, usually referred to,

Paper replaces the wing-root fairing and the spinner shows signs of severe damage. The paint job is highly imaginative!

erroneously, as a 'Galland Hood' was a belated attempt by the Erla facility to improve visibility from the cockpit. Most of the heavy steel framing of the original design was eliminated, as was the rear, jettisonable glazing, and the whole replaced by a single unit. It made an appearance on late-production Bf 109 G-6s, but was certainly never fitted to the G-2.) In some trepidation, I balanced precariously on the in-built fuselage step and surveyed the scene. It could best be described as a disaster area! Readily noticeable, the instru-

Left *The cockpit in 1962. Vandalized but reasonably intact.* **Right** *The same cockpit ten years later! On the left, above the priming tank can be seen the Meteor fuel cock which had been fitted to serve as a throttle.*

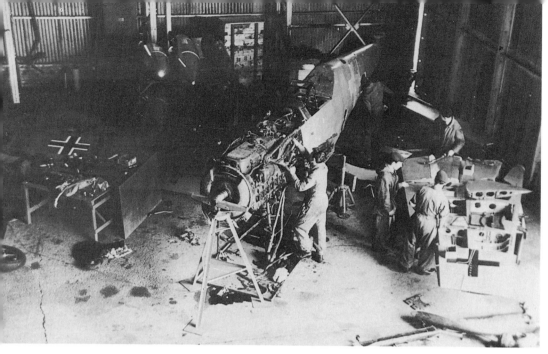

Wattisham 1962. The 109 has been dismantled for an abortive attempt to make it airworthy.

ment panelling had been replaced by a crude piece of aluminium alloy sheeting cut to display a minimal array of instruments, most of British origin. The complicated electrics panel on the starboard wall had vanished, and the

The lower cowl hangs from a very untidy engine.

throttle quadrant modified to accept a lever externally. This was later found to have been a fuel cock from a Meteor! And everywhere lay loose cables, nuts and bolts and an amazing depth of dirt.

The following day, refreshed by a good night's sleep (induced by a modest intake of ale), I set about an in-depth examination of the hulk and, with the help of some curious bystanders, I removed the loose upper engine cowlings. Most of the fasteners had been mangled and the skin around them torn and bent. At the rear, the rubber sealing strip had all but vanished. On top, where the muzzles of two machine guns once protruded through oval apertures, pieces of alloy sheet had been crudely spot-welded. The lower cowling, which is hinged on its right edge, was carefully unfastened and lowered. The dormant weight of the brass and copper oil cooler contained therein had obviously been the innocent cause of severe damage. Once again, the retaining fasteners had been mutilated and the hinge assembly badly damaged. The fairing into which the cooler was inserted had also suffered, and several disturbing cracks in the skin were discerned. The engine itself could best be described as 'tatty'. Fortunately, however, it was still protected by an ancient coat of preservative. The top cover had lost most of the retaining nuts from its forward end, and some ignoramus had bent the hefty panel upwards, presumably in an effort to view the interior. The annular oil tank which conceals the propeller reduction gear at the front end of the engine had a nasty large dent on its lower periphery. Underneath, the fuel injector pump appeared intact, but its pipework, together with most of that of the engine, had been damaged or was missing. Forward of this lay two, omi-

A view showing the lower engine installation which was of great use during the rebuild.

An engine strip begins at Wattisham.

nously empty, clamps. Formerly they had held the electric propeller pitch control motor. To the rear of the impressively large Daimler-Benz engine, copious bare-ended wiring clearly indicated missing electric plugs and connectors.

The fuselage was hideously clad in a coat of gloss, sand-coloured paint. This did little to disguise damage to the skin. On the panels which formed the cockpit walls, areas of scoring were found. (Initially inexplicable, this damage was found subsequently to have been caused by inept handling during the movement of the aircraft on its many road journeys. The mainplanes incorporate heavy steel lugs bolted to the spar, through which pins are inserted to mate them to the fuselage. On the 'Queen Mary' transporter, these had been laid on their leading edges and propped against the fuselage, the protruding fitments transmitting a substantial proportion of the weight to small areas of alloy skin.) Small dents were to be found everywhere, and underneath near the tail-leg a more severe area of damage was apparent. The rudder was badly deformed, although still completely fabric-clad. The upper hinge mounting within the fin had also been damaged and it seemed that the aircraft had been pushed rearwards into an immovable object.

I next removed the radio hatch, itself bearing damage commensurate with having been prised open. Inside could be seen a large metal label describing the radio equipment. Someone had attempted to remove this, one corner being badly bent around a retaining rivet. Forward, the tubular radio racking was still in situ, but the equipment itself was missing. Indeed, little remained in the rear fuselage, bar the distinctive smell found in all German aircraft of the period. (I have heard many explanations of this musty odour,

On trestles, 10639 has been shorn of all its markings.

but I feel that it was almost certainly because of the synthetic fuel and oils employed by the Luftwaffe.)

The internal inspection took far less time than I had expected (hoped!) and, with sinking heart, I turned my attention to the undercarriage. The tail-leg seemed intact but it carried a very small wheel which I guessed — cor-

An intact but grubby DB 605. The belly panel hanging below has been fitted back to front!

rectly — was a Spitfire unit. The main legs bore two authentic but dissimilar wheels, both of which were damaged. The tyres were in perished condition and proved to be agricultural covers with the deep tread machined off. Although the brake units were in place, all the system piping had vanished. Atop the legs, crude pieces of iron had been inserted as makeshift down-locks.

The mainplanes did not provide any source of comfort either. Minor damage was to be found everywhere, including holes drilled through the thick gauge skin on the inner surfaces as a means of attaching the cardboard fairings. At the trailing edge, the radiator flaps had been secured in the closed position using pop rivets, and the plain flaps had both received extensive damage. The ailerons seemed structurally intact, but both detachable wing-tips had been scored, and seams had been torn apart. The pitot head assembly (the air source for several of the cockpit instruments) was missing, and within the wheel wells lay the remnants of one of the two canvas screens which were intended to exclude from the wing interior any foreign matter dislodged from the wheels on retraction. One hydraulic actuating cylinder for undercarriage operation had been pilfered, and everywhere panelling was 'secured' using a minimal number of screws, most of which appeared to be of an incorrect type.

Finally, I examined the propeller which seemed in reasonable order, although the blade tips were slightly bent and all three bore minor damage. The spinner, on the other hand, was a sad sight. A distinctive feature of the later 109, the tip of the streamlined fairing had once been formed to permit the exit of ordnance from the engine-mounted 20mm cannon. A large, crude saucer-shaped disc had been riveted thereon, and most of the remainder of the original unit had been badly abused.

The overall picture, therefore, was one of dereliction, mistreatment and neglect. I suppose I should not have been surprised as I had seen several other German aircraft, all in similar condition. What worried me most was the missing equipment. It would all have to be replaced, of course, but I had no idea where I might find it. I had seen nothing in the damage that deterred me. Although it had all been caused by unprofessional handling and lack of care, it was generally minor and I estimated (*slightly* optimistically, as it turned out) that five years would complete the project.

One essential component of the grand plan was missing, however. As a pilot, I was not admirably qualified to rebuild an aeroplane! I desperately needed good engineers — good, volunteer engineers. Within days, the first man stepped forward in the shape of Chief Technician 'Paddy' Stanbury. Fresh from months of work on exhibits for the RAF Museum, and within months of retirement, he took to the 109 immediately. One of my fondest memories of his, albeit brief, involvement was the afternoon we removed the propeller. Digging into his long experience, he decided to employ a 'Giraffe' for the task, there being no suitable lifting gear available. The 'Giraffe', I should explain, is a tall, mobile servicing platform which can be raised and lowered using a basic hydraulic system pressurized by handpump. Paddy elevated the platform and positioned it above the propeller boss. Grabbing a roll of fabric lashing tape, he attached a length to each of the two uppermost blades and secured the other ends to the platform framing.

With the weight being taken by the rig, we eased the prop from its shaft, and it was then the fun started! As I pulled the lower blade toward me, Paddy lowered the platform, the idea being to lay the assembly flat on the hangar floor. It descended with frightening speed, particularly as I discovered that it was far too heavy to guide into the required horizontal position. A young airman passing by came to my rescue, but even his muscle power was unable to improve the situation significantly. Paddy was not impressed, claiming that he had performed this same task on Spitfires on many occasions. Somehow we managed to lay the precious item gently on the floor, Paddy all the while pouring scorn on our physical abilities. Striding to the prop hub, he decided to demonstrate how easy it was to lift. After a few seconds of mighty effort, during which no movement was noticed by we lesser mortals, and culminating in much heavy breathing and a very flushed face, he then began to realize his mistake. The 109 carried an extremely weighty propeller!

Hot on the heels of Paddy came Sergeant Frank Shaw. Frank was considerably younger than Paddy and had never worked on an old aircraft, a background he had in common with all those who followed. On reflection, I cannot claim that he exactly begged to be allowed to help. Indeed, apart from the odd question on the aircraft, he showed little inclination to offer practical help. I was naturally anxious to gather a team around me, and on this single occasion I bent my own rules and subjected him to a programme of gentle persuasion. I hope that he has long since forgiven me!

Within a few days, the engine was removed from the airframe for the first time since the Second World War. The next step was a mystery, and a worry. There was no-one at Lyneham qualified on heavy piston engines, and certainly no-one who had any knowledge of the Daimler-Benz DB 605 A. External help had to be found. The Daimler-Benz company had not built an

View of front fuselage after engine removal. At the bottom lies the coolant thermostat.

One dirty and incomplete engine. The sump cover may be seen at the top, front end bent upwards.

engine of this class since 1945 and were only likely to be able to advise. (A forlorn hope, it transpired.) Not far away, at Bristol Filton, was Rolls-Royce, of course, but I was convinced that an approach for help from that quarter would prove fruitless.

In HAPS days, we had been offered, free-of-charge, a Fairey Firefly Mk 1 by Svensk Flygjanst AB. The company had used this venerable aircraft to tow targets for the Royal Swedish Air Force but it had just been withdrawn from service and replaced by reworked, ex-Royal Navy, Douglas Skyraiders, by a curious quirk of fate the same aircraft I had seen lying at Abbotsinch years before. Prior to permitting a ferry flight to England, however, a check was needed of the Rolls-Royce Griffon engine and our Swedish benefactors informed us that this could be arranged, but at a cost of £150 — the fee of the proposed engineer. I wrote R-R (albeit Derby) requesting a contribution or, failing that, the required examination. The rebuff came as a bitter shock and, because of the time constraints placed upon us by the Swedes, precluded any further search for help. We lost the rare Firefly and it was subsequently scrapped. As a result of this experience, I was disinclined to approach R-R again, particularly in view of the precarious state of the company at that time.

The Public Relations Officer at Lyneham made a suggestion which I rejected initially. He proposed organizing a television interview which could be broadcast in the form of an appeal. I eventually relented and, within a matter of days, a team from BBC Bristol arrived. Some hours were spent filming and, that evening, the interview was broadcast on the regional 'Points West' news programme. Much to my surprise, there was a speedy response and from the most unlikely (in my opinion) source — Rolls-Royce Bristol! Days later a delegation arrived and expressed its delight in the condition of the engine. Could I make arrangements to move it to Filton?! It transpired that the Apprentice Training Department had hoped to rebuild a Bristol Jupiter engine which would then have powered a Bristol Bulldog fighter owned by the Finnish Air Force Museum, but lines had somehow crossed and the plan fell asunder when it became evident that the Finns had not intended to fly the aircraft at all. Bristol had been unable to find another project which would result in a rebuilt Bristol engine being flown, and the Training Department had decided to widen the field and seek an alternative project which required the rebuild of a large piston engine. The interview on 'Points West', therefore, was opportune. The engine left Lyneham a week later for an overhaul upon which no time limit was placed, but which R-R estimated might require three years. Amusingly, some weeks previously I had discovered a flight engineer from the Hercules fleet poking around the engine. I told him that I was hoping to find a company willing to take it in hand. Displaying a remarkable degree of self-confidence, he retorted that, had I given it to him, he would have had it running within six months! It seems that my ability to underestimate the magnitude of a task was not unique.

With little direction from me, work on the airframe began in sporadic fashion. Paddy found some time during the working day to repair the damage found forward of the tail leg. Frank, however, was in charge of the very

Aircraft laid out for an official inspection, paint removed from all external surfaces.

busy Structures Bay and was unable to assist in similar fashion. We decided to devote each Sunday to the 109 and, once we had removed the fin from the fuselage, set about repairs to the rudder attachment in the relative comfort of his workshop. Pleasing though it was to have these first repairs effected, it was important that work began on removing coats of modern paintwork. By then, I had received photographs taken of the aircraft at RAF Wattisham, Suffolk in 1962 which clearly showed, much to my intense disgust, that all earlier paint (including the original German) had been removed. Hence, there was little to be gained in carefully removing existing paint layers as, on the external surfaces at least, nothing remained worth recording.

I found little enthusiasm when I described my plans to my workmates and I decided I would tackle the stripping myself, thus allowing the continuation of repair work. I realized then that the interest of Paddy and Frank lay more in the engineering challenge of the project; neither had any inclination to be involved in research and restoration. This was a blow to me, and I could foresee a conflict of interests in the future. My disappointment was short-lived. Unbeknown to me, one of the men who had removed the aircraft from the Customs Shed on the day of its arrival had approached Frank Shaw to ask how he could become involved. His name was Ian Mason, then a corporal in charge of a small team on the airframe side of Comet engineering. To this day I consider myself a very lucky man that he stepped forward. He immediately displayed an uncommon enthusiasm for all aspects of the project, and, in the years that followed, was to prove instrumental in its continuation. He joined me in the messy, unrewarding drudgery of paint stripping and, within only a few work periods, the fuselage and both mainplanes had been shorn of their pathetic psuedo-camouflage.

During this process, besides grabbing the moments we could during the working day, Ian and I agreed to spend Thursday evenings in the hangar. One such evening, we both reported for a few hours work, only to discover that the 109 was not where we had left it that afternoon. We discovered it in a far corner. Not only that, but all our cleaning gear had vanished, effectively cancelling our night's toil. The following morning I strode into the office of the Squadron Leader engineer to protest. He professed to know nothing about it and summoned his Flight Sergeant, who blatantly admitted that he had decided to reorganize 'his' floor space. In some rage, I tore him off a strip, pointing out that, as he knew nothing of our work, it was conceivable that his actions might have caused further damage. I forbade him touching the aircraft unless he first consulted me, and at least one of my colleagues was in attendance. This was accepted in scarcely disguised bad grace, but the episode provided an insight into the attitude of the engineering staff.

Although, prior to the arrival of the 109, I had consulted all my superiors for permission and hangar space had been granted, it was apparent that the engineers in charge of the hangar resented the appearance of a derelict machine in their otherwise tidy building. Further, from comments made, I deduced that they were more than a little unhappy that a pilot was dabbling in matters engineering. Thankfully, however, there were many others with opposing views and they made life tolerable, for it was a situation with which we were to become very familiar during the span of the project.

Another familiar situation arose shortly after. Frank Shaw was promoted and posted to Swanton Morley in Norfolk! This move was totally unexpected and curtailed an extremely valuable channel of help. Prior to this, Paddy's contribution had ebbed away as he anticipated his retirement, and all progress then became dependent on Ian and myself. Our main effort focused on the continuing strip of the fuselage, the rear portion of which is remarkably small. We established that the only way of entering was through the fuel tank bay, but this first involved the removal of the rubber tank mounted behind and below the cockpit. Conscious of its susceptibility to damage, we took great care in extracting it. A survey revealed it to be in remarkably good condition, although lacking various parts of fitted equipment, and a short while later we had it despatched to RAF St Athan in South Wales to be inspected and tested by the specialist fuel tank cell there.

One barrier remained, preventing our entry to the rear fuselage. This was a vertical screen of armour protection. Two halves, shaped to accommodate the passage of the fuselage stringers, were joined along the centre line by two heavy gauge aluminium-alloy plates and fastened by numerous nuts and bolts. Once removed, we were astonished to find that it comprised 27 sheets of alloy sheeting pinned together, affording a total thickness of 20 millimetres!

The fuselage was then rapidly stripped of wiring and hydraulic, oxygen and compressed air pipes, plus the radio racking, and Ian and I set about the lengthy task of removing years of accumulated filth, frame by frame. As only one man could fit into the fuselage, we performed the cleaning in shifts, taking great care where we applied our respective body weights. This tedious work took months to accomplish and became even slower as we penetrated to the extreme rear.

Inevitably, as no change could be discerned by the passer-by during this period, we found ourselves continually explaining our work and protesting that we had not given up! It was galling for us to read years later of the equipment available to the staff of the Smithsonian Institute, which they employed to clean the skin of their Messerschmitt Me 262 A jet fighter. Propelled by compressed air, crushed walnut shells bombarded the surfaces and left them spotlessly clean and devoid of surface corrosion. Although this was, by all accounts, not a pleasant task, Ian and I both consider it far preferable to the uncomfortable months spent in the cramped rear fuselage of the 109, wielding brush and fine grade wet and dry abrasive paper, and slowly removing our finger-tip skin! Given the use of such specialized equipment, I hate to reflect how swift would have been our progress.

Not only had we to take extreme care where our weight was applied on the fuselage 'floor', but we had also to position a paper cup containing thinners with which to dissolve the grime and reveal the grey-green primer paint. This we brushed on small areas of skinning. Because of the confined conditions, we soon tired. On one occasion, Ian buried inside, myself nearby, I was startled by a loud, muffled — but readily intelligible — oath. The fuselage shook alarmingly on its stands and appeared to be in real danger of toppling. A cup full of thinners had been upset and the contents had run down the keel and eventually poured out through adjacent drain-holes. The violent movement of the airframe was a result of Ian attempting to levitate his not-inconsiderable bulk away from the flow!

Whilst we laboured during the day, I used my evenings to explore the possibilities of assistance. I began typing letters to companies, both at home and abroad, whose names I found in the Farnborough Air Show Trade Directory. Eventually, the format of a standard letter evolved which, with minor adjustments, became my plea for help. The responses were varied. The occasional expression of interest was, all too often, interspersed with flat rejection, usually accompanied by explanation and apology. Frequently companies deigned not to reply at all. I contacted Dowty-Rotol regarding our VDM propeller, but they declined to help. A few weeks later, British Aerospace Dynamics of Lostock agreed to overhaul the assembly.

At around the same time I was invited by the Triplex company to bring the damaged armour glass and cockpit plexiglass to Birmingham. I arranged the use of a Service van, proposing first to travel to Birmingham and then on to Lostock to deliver the propeller.

Our first problem came early that day when Ian and I discovered that we could not quite fit the prop into the van. (I destroyed my favourite jacket in the process, a seam ripping apart on the back whilst I struggled to jockey the awkward load.) Abandoning the assembly (which was subsequently delivered in the company of four others from Lancaster PA474) we drove to Birmingham. The Triplex executive greeted us at the reception desk, and our conversation took place there, amidst considerable hubbub and turmoil. It was apparent that we were not sufficiently important to be offered a seat in a quiet office, and it set the tone for what followed. We showed him the armour glass. He explained that the company had been experiencing hard times. Had we come a year earlier .... As it was, they could not afford to devote the time or resources to reproduce the piece. 'Anyway, why not make it yourselves? Try glueing a few sheets of plexiglass together and then shape them'!!! We then turned to the various small windows of plexiglass. Most of their edges were rebated, whilst three bore flanges. Again the suggestion was that we make them ourselves, and with that the interview terminated. We were appalled at such offhand treatment, and the memory of the incident rankles to this day. After all, having been invited to Birmingham, not unnaturally we had reason to hope an offer of assistance from this large company would result.

1974 was drawing to a close amidst much speculation on how a further round of defence cuts would affect 216 Squadron. Opinions varied, but there was an all-pervading and overwhelming feeling of doom. At best we felt that the Comet would continue flying in the VIP role for some time, as the alternative for the task was the much larger and expensive-to-operate Vickers VC10. We also thought that the elderly Bristol Britannias would receive the axe first. But our worst fears were realized just prior to the public announcement of the cuts: Two-Sixteen was to cease operations within a matter of weeks.

Many highly experienced aircrew, disillusioned, took advantage of financial inducements and resigned their commissions. Most readily found employment in the airlines, but I was then of an age which attracted no cash offer to leave, and I was obliged to await a posting. Threatened with a non-flying job (as were most of my remaining colleagues) my boss took a hand in matters. Fortunately, it seemed that I had fulfilled my role in 216 adequately and, as I had held a command on the Comet for only a few months, he urged that I be allowed to continue VIP flying.

Just before I set out across the Atlantic on my last tasked flight, he told me that I was to take up a a post with 32 Squadron flying HS125 executive jets at RAF Northolt. My elation at the news lasted only until my return, for it was then I learned that Ian Mason had been posted to Brize Norton. What was to be done with the 109?

4

Move to Northolt

Faced with the uncertainties caused by the demise of Two-Sixteen, work on the aeroplane ground to a halt. It was obvious that, whatever the future held, many months would elapse before it could resume. My first priority was to repeat the process of seeking hangar space on my new station, particularly as I was aware that Northolt, although a small airfield, was home to a large fleet of aircraft. It was quite possible that room could not be spared. However, my luck held and a telephone call confirmed that the little fighter would be shoe-horned in, somehow.

While Ian and I set about preparing the myriad parts (large and small) for transportation, our mood was one of despair. After all, we had just got to grips with our voluntary task when fate conspired to place a large spanner in the works. It seemed an awful waste of valuable time — packing when we could have been continuing the strip of the fuselage. However, day by day more parts became enwrapped in plastic sheeting or disappeared in cardboard boxes. Packing materials were abundant; the Squadron inventory was being dismembered as equipment was transferred to other units. Transport, too, proved easily come by, but inevitably we had little choice when it came to timing the move of the aeroplane. By then, I had started my conversion on to the HS125 and Ian was left to supervise both the loading and the move. It all went remarkably smoothly, and a large number of packages were deposited in the smallest of Northolt's three hangars, there to rest unattended for some time.

I had wondered how Ian felt about his posting to Brize Norton. His new job would take him away from engineering proper and place him in the role bay, where his duties would revolve around preparing the equipment and furnishings required to be inserted into transport aircraft to enable them to fulfil their allotted tasks. He was pretty despondent at the prospect. My second question probed the depth of his interest in the 109. Specifically, I was curious to learn how he felt about Northolt as an alternative posting! I was delighted — and relieved — when he reacted positively to the suggestion. It was only wishful thinking at the time, of course, but he gave me his blessing to probe the possibilities. I foresaw little point in approaching the postings branch. Every year in the Service, my performance was assessed, and the relevant form invited me to submit my choice of posting. Suffice it to say, my preferences were religiously ignored! Besides, a mere Flight Lieutenant had little authority, but perhaps the Chief of Air Staff had.

Some months previously, we had received a visit from Air Chief Marshal Sir Andrew Humphrey, who had expressed a great deal of interest in our work. On taking his leave, he offered (rather rashly I thought!) to help in any way that he could, should I encounter problems. The gesture was much appreciated, but I doubted that I would ever have need of assistance from such high level. More in hope than conviction, I penned a letter to his office. Some weeks later, reaction came in the form of several phone calls from Whitehall. Sir Andrew had directed that moving Ian to Northolt was to be investigated, but it was thought unlikely. And, in any case, how did I know that he wanted such a move?! I resisted the sore temptation to respond to this ludicrous question, suggesting instead that it be put to him directly. I began to wonder just how much power CAS actually wielded, and was almost sure that I had wasted my time. But Ian was asked, and soon learned that his days at Brize were numbered and that he would be on his way to north west London.

My flying career resumed in October 1975 upon completion of the required technical courses. Whilst learning to fly the delightful little executive jet (used by 32 Squadron as a VIP transport), there was little available time for the 109 other than to ensure its safety. Ian's arrival in January provided the impetus to restart the project, but first the aircraft had to be moved into one of the larger hangars where we had been allocated the perfect corner; it was but a few seconds walk from my Squadron's offices! But then the first problem surfaced. During my earlier negotiations for hangar space, I had broached the subject of shelving for our equipment with the Northolt supply officer, and had been assured that some would be made available. When I requested it, however, I was told bluntly that none could be spared. On the other hand, while I found little sympathy for my predicament, great concern was expressed that my boxes were rendering the hangar untidy! By chance, I discovered that some ancient iron shelving had been discarded some months before, and I was permitted to retrieve this, if it could be found! Ian and I, plus personnel of the Visiting Aircraft Servicing Flight (VASF) spent a number of days extricating rusty angle iron from thick undergrowth. This we painted (I daren't risk being further accused of making the hangar untidy!) and then scrounged bolts with which to assemble the Meccano set. Only then could all the aircraft components and tools be unpacked and the project resumed. Over nine months had been lost.

During this period, I had not neglected a very essential ingredient of any restoration — the search for information. The stripdown of the 109, and most repairs, were relatively straightforward. However, sooner or later, work would have come to a halt because of the lack of manuals. I made an appointment with the Documents Department of the Imperial War Museum in Lambeth. On arriving I was allocated a table, and I set about the search.

First sight of an identity plate. Note the Erla inspection stamp to the right.

Within minutes I had stumbled upon the first reference, a manual entitled 'D.Luft.T.2109 G-2', and I asked for the relevant role of microfilm. It was just what I needed — the technical manual for the Bf 109! Some of the lesser sections were not held, unfortunately, whilst others had yet to be transferred onto microfilm, and these I was also allowed to examine. I completed an order for photocopying, only to learn that it would be over two months before I could hope to receive anything because of the backlog of orders! Worse, the cost per page meant that my eventual requirement would be extraordinarily expensive. I appealed to IWM to aid the project, suggesting that the photocopying could be provided more cheaply, and where two copies of a manual section were held, one could be released into my charge for copying elsewhere. Not unexpectedly, the Director flatly refused and I faced my first large, personal expense.

Eventually, with the aeroplane arranged neatly, in kit form, on the allocated floor space, we were ready to recommence the stripping of the fuselage. It was the turn of the front end, and Ian and I carefully removed the windshield, followed by all engine controls and electrics from the front bulkhead. This permitted the dismantling of the complex structure forward of the cockpit which serves as a fire bulkhead, although a poor one. Once removed, we found the first positive proof of the mark and origin of the aeroplane; a small alloy label clearly identified it as a Bf 109 G-2, built by Erla Maschinenwerk GmbH, Leipzig, N.24! It had been totally immersed in a thick layer of dirt and engine oil. Finding it gave me a little thrill, and I understood why archaeoligists endure endless toil with trowel and brush.

At the same time, though, we made a discovery which sowed a seed of suspicion that ours was not a standard production airframe. We found a curious modification to a lever which formed a small link in the engine throttle control. The complete linkage had been anodized (a process which coats aluminium alloy with a thin layer of aluminium oxide by electrolysis) and, in common with all other components, was in *beautiful* condition. I commented at the time that it was a great pity that the whole aeroplane had not been thus finished! But I digress! This lever had a circular hole cut in it and stamped next to it was 'E u N'. A similar hole had been more crudely cut in such a fashion that the two perimeters overlapped, and this was identified as 'A-motor'. The explanation of this was that the earlier Bf 109 F had been powered by the Daimler-Benz DB 601 E and N (E und N) engines. The Bf 109 G was equipped with the later DB 605 A. This component, therefore, was made for an F-series aircraft but had been modified to G-series throttle configuration.

Sergeant Ian Mason (promotion came during his move to Brize) spread his enthusiasm for the project to his men on VASF. Three were sufficiently motivated to begin helping us, and this boosted our efforts considerably. We removed everything from the cockpit and carefully bagged each item complete with descriptive labels, and I then embarked upon the necessarily slow process of carefully lifting paint which had been applied since the War. The whole area was a mess. Large quantities of paint had been slapped on by brush, layer upon layer. Indeed, so much paint was there, that it took some time to locate and uncover various metal labels, including a large one on the port wall. We had no idea that it existed; it had been totally camouflaged by black paint! Hand-painted instructions for the tailwheel lock and a decal for the battery switch came to light and were recorded on film before erasure. It was important that this was done in order that everything could be reproduced. Suffice it to say, it took several weeks to complete this work and also remove an amazing depth of accumulated dirt and sand from the cockpit floor.

At that time we still lacked German manuals to aid the dismantling, as I had had no success in finding a willing translator. Extracting the control linkages above and below the floor by trial and error proved difficult, and a lot of blood was shed when spanners and screwdrivers slipped, or when skin was removed from hands and arms inserted into tiny access holes. German engineering became the butt of much adverse comment! I believe we all found the language interesting, too! I can claim, without any fear of contradiction, that I learned more Westmorland vernacular than Ian did Scots, although subsequent years served to redress the balance — slightly! Oaths were usually accompanied by calls of: 'Don't you bleed on my aeroplane!', blood being a highly corrosive fluid!

Removing the complex and awkward rudder assembly and the 'tray' which forms the base of the engine-mounted cannon cover allowed the unbolting of the forward cockpit floor, and stripping and cleaning accelerated dramatically. Meanwhile, our new team members, Kevin Thomas, Pete Hayward and Tony Leek, devoted their time to cleaning components and helping Ian repair areas of damage to the 'box frame' (as the structure forward of the cockpit became known). Pete, an electrician by trade, turned his attention to servicing the components which existed, and listing those which didn't. There were an awful lot of these, and a few months later, we were to discover the true extent of the problem.

I was extremely worried by the lengthy list of missing equipment in general; it was far more serious than my initial survey had suggested. The RAF Museum store at Henlow provided two or three instruments and a *Mutterkompass* (the detector unit of the compass system), but I needed a lot more and began investigating potential sources of spares. The most obvious was the Finnish Air Force, which had been the last operators of the Bf 109 G. Finally withdrawn from service in 1953, two examples were on display in Finland. Jack Bruce of the RAF Museum had been singularly unsuccessful in dealing with the Finns, but by a stroke of good fortune, I established a line of communication through another well-known figure in the historic aircraft field, the late Doug Bianchi of Personal Plane Services, Booker Airfield. During a visit he allowed me to browse through the hangars, and even spared time for a chat. When I mentioned my search for spares and my desire to contact the Finns, he immediately suggested Major Kauko Rasanen. He had found this officer most helpful following the acquisition of a Vlima biplane trainer. Consequently, I wrote him at Finnish Air Force Headquarters at Tikkakoski and was delighted to receive a prompt response, but not surprised when it carried a 'shopping list' of parts for the Finnish Air Force Museum! As they claimed to hold some 20% of the parts I had listed for the 109, I was spurred into finding something to offer in exchange. The largest item needed was a Rolls-Royce Merlin Mk25 engine for their Hawker Hurricane, and, over a period of some months, I tackled all the preservation societies, museums, companies and individuals of my acquaintance in the hope of finding one, but without success.

My last hope was the Battle of Britain Memorial Flight. I was doubtful that they could help, and was amazed to learn that they actually held in store a Merlin of the correct mark! Not only that, but as it had a dubious history, it was of no practical use to the Flight, all useable parts having been removed. It transpired that Ford apprentices had assembled the engine from a motley collection of parts as a training exercise, and then presented the result to BBMF. My contact reasoned that I stood to gain a great deal for the 109 and immediately suggested that I offer the Merlin. I proposed putting on paper the transfer to me to make it formal, but was deterred from so doing because of the threat and likelihood of long bureaucratic delays ensuing. This was a *major* error. However, an exchange was agreed with the Finns and I expected to receive some instruments, a spare canopy, wing-root fillet and belly panelling. The deal, though, was dependent upon me arranging transport to and from Finland! I approached several freight companies and received negative responses, until I spoke with a Philip Mann, of Mann and Son, whose eagerness to help was refreshing. He, in turn, personally arranged not only the shipment of the engine to Finland, but the return passage of the exchange equipment, free-of-charge, through the Bore Steamship Company.

The next step proved extremely expensive. I could not locate a suitable crate for the Merlin, and, regrettably, had to appeal for RAF Museum help. Always willing, an extremely costly box was manufactured at Hendon. Everything arranged, I called BBMF, only to discover that my contact had left (the Service!) and I could find no-one who felt able to release the engine! I telephoned Strike Command Headquarters and spoke to a Squadron Leader who held overall authority for the engineering of BBMF. I detected immediately that he was far from sympathetic, and within a space of a few minutes I was aware that he was downright hostile! Not deterred, I pursued my aim to have the 'useless' engine released, but this officer was not prepared to listen to, or accept, my account of the events which preceded the call. A few days later, having been unable to make headway in other directions, I called him again. This time, in the course of a fairly heated discussion, he claimed that the engine had been disposed of. When he accused me of dishonesty I curtailed the call abruptly. Somehow, I had to bypass him if I were ever to succeed in extracting the engine. I decided to ask Jack Bruce to write an appeal to a more senior officer at Strike Command for co-operation in the matter, as it was likely that liaison with the Finns was liable to benefit not only the 109, but the Museum as well. I subsequently received a letter which very reluctantly agreed to release the engine.

Whilst I was very glad that my ploy had succeeded, I was more than a little upset by a blatant statement in the letter which declared that he (a Group Captain) did not believe that any of his staff had agreed to surrender the engine. I saw red and was inclined to demand an apology. However, there was a problem. Firstly, the suggested transfer of the engine was never put in writing and, therefore, I could not substantiate it without drawing my earlier contact into an acrimonious argument, possibly causing him a great deal of trouble, and this I had no wish to do. (I had no idea whether he had broached the matter with his superiors.) Nevertheless, I was very hurt by the accusations of these two officers and, to this day, I have great difficulty reconciling their behaviour with their status. Most rankling of all, I had to submit to this abuse, not for personal gain, but to improve the condition of an aircraft which was the property of the Ministry of Defence. Yet I had been treated as an 'enemy'.

The trade with the Finns was accomplished, not without minor problems, and my liaison with them developed nicely. One of its staff, Pertti Virtanen, was scheduled to visit the Farnborough Air Show that September and he requested to see our 'baby'. Ian and I met him at his London hotel and there followed a fruitful, yet tiring day. Pertti could speak no English and, somehow, Finnish had been overlooked in my education! His Embassy had arranged the presence of one of its secretaries as ad hoc interpreter and, whilst we were thus able to converse, we discovered that the translation of aeronautical terms was beyond her grasp. Pertti had brought with him the *Ersatzteillisten* (spare parts manuals) for the Bf 109 F-1 to 4 and the Bf 109 G-1 to 4, and whilst he surveyed the aeroplane, Ian and I browsed through the pages. I had believed these tomes superfluous to our work, but examination of the hundreds of pages confounded that view; they were amazingly detailed and instructive. Whilst we were not permitted to retain these books, we were given one night in which to photocopy them. Many hours were needed during the course of that evening for the two of us to reproduce the volumes which were to prove by far the most valuable components of our expanding library. I now feel able to confess that, in so doing, we rendered two photocopying machines unserviceable and caused no little inconvenience the following day to at least two offices!

During several months thereafter my new team developed a work pattern which, although hardly regular, achieved significant advances in the process of clean and repair, and I began to anticipate the start of the rebuild. At the same time (early 1977) Northolt was scheduled to close its runway for several months whilst it was resurfaced. The resident Squadrons learned that their aircraft would be dispersed to other airfields for that period, with the HS 125s operating from nearby Heathrow. I foresaw little in this arrangement which would affect the project other than to allow us more free time! My optimism proved short-lived. One day, whilst labouring in the hangar, a corporal innocently asked when I was moving the aircraft. I denied that I was, but he was adamant that a move had been decided by his boss. As I had not been consulted or informed I concluded his assertion to be hearsay. Two weeks later, though, the Station Engineering Officer casually confirmed the move and suggested that it had to be completed within the following two days! I was flabbergasted, particularly as he had seen fit to let the whole hangar know before the man in charge.

Worse followed. Our new work space was to be the corner of a dilapidated Nissen hut! The move was effected during one working day, with the help of a considerable number of engineering personnel, and the aeroplane was deposited inside the doors at one end of the shed. I had imagined that this relocation had been necessary because work was scheduled on the hangar whilst the airfield was devoid of aircraft. In fact, the entire hangar stood empty for months whilst we languished in a dismal storage building. Needless to say, I was very angry.

I would be less than honest if I were to deny that my opinion of the engineering officer corps had been untarnished by the events I have related. However, if it had not been obvious at Lyneham, we knew that we were very much 'on our own'.

5

Friedrich or *Gustav*

I had yet to overcome the formidable barrier presented by the German language. Neither I, nor any of my team, had more than a rudimentary grasp of it, and none of us could aspire to the production of a translation of the servicing manual in all its parts. Always the optimist, I approached a professional translator of technical German. I had no doubt that discussing a fee would be a pointless exercise, but I hoped that an appeal for help might prove fruitful. Not for the last time, I was to be disappointed. The cost per thousand words of translation would have consumed a fair proportion of my RAF salary!

My thoughts then turned to the Federal Republic of Germany, within whose boundaries lay many British military bases, including several RAF stations. So, I composed an appeal which I submitted to the education service for transmission to their officers. Sad to relate, the eventual reply was a negative one, but attached to it was a copy of my letter which bore evidence of having been read by several of the intended 'targets'. 'Suggest Snadden finds a professional translator.' 'Does Snadden realize how much such translation would cost?'!

Fortunately, help was at hand. Following publication of a letter in *Flight International* magazine which included a request for the donation of parts, I received a letter from Peter Nolte in Bremen. Over the ensuing months I was able to meet him and discuss the needs of the project. As it was my priority at the time, I broached the subject of translation and was delighted when he immediately proposed that he do it himself. Selecting the potentially most useful part of the *Handbuch,* I sent it to him, and so began a process which was to cover months of his spare time and provide me with a well-nigh complete library — in English. Peter proved to be a real asset to the restoration, finding me many components and arranging German assistance.

By then Ian and I had read just about every book written on the Bf 109. We discovered many anomalies and it was impossible to know which author was correct. Photographs were particularly aggravating. It is astonishing how many were (and are) incorrectly identified. The picture became much more clear to us when the handbooks divulged their information, and the hundreds of pages of illustrations in the spare parts manual had been absorbed.

It is commonplace for photographs depicting the Bf 109 F, for example, to be misidentified as the Bf 109 G, and vice versa. It is an error easily made, as the two marks are very similar, but it is also an error easily avoided with a dusting of knowledge. The F-series, often referred to as the *Friedrich* from the German phonetic alphabet of the day, first saw service towards the end of 1940, a scant few weeks after the Battle of Britain. It was a radical improvement on the earlier Bf 109 E. In particular, the angular engine cowlings of the *Emil* gave way to simplified panelling. The new nose appearance was aesthetically most pleasing and a great advance aerodynamically. It was graced by a propeller spinner larger than its predecessor but in perfect proportion. The tailplane was devoid of bracing struts for the first time, and the tail-leg was made retractable. The mainplanes had been redesigned and incorporated small leading edge slats and a revised flap configuration. Most obviously they contained no internal armament and bore extended, rounded wingtips.

Six production versions of the *Friedrich* were built, but only three in real numbers. Distinguishing one sub-series from another in photographs is difficult. The first new-shape aircraft was the Bf 109 F-0, a pre-production machine of which there were ten. It is most easily identified as it sported an angular supercharger intake on the left cowling, reminiscent of that used by the earlier Bf 109 E. The armament consisted of two MG 17 machine guns ahead of the windshield and an Oerlikon MG FF cannon of 20 millimetre calibre firing through the spinner.

The first production model, the Bf 109 F-1, was identical in every respect, except that the supercharger intake had evolved into a more pleasing circular-intake form, with a narrow 'lip'. Following structural failures of the tail unit, the area was strengthened by the fitment of four external straps over frame 9 of the fuselage, under the tailplane.

Confusingly, the early production Bf 109 F-2 also had its tailcone similarly reinforced, but later aircraft benefited instead from internal bolstering. Cannon equipment, however, became the faster-firing Mauser MG 151/15 (15mm calibre) and in some nose photographs, it may just be possible to discern a small muzzle diameter in a comparatively large spinner orifice. To muddy the waters even more, late production F-2s had the MG 151/15 replaced by the much more effective MG 151/20! Thus, the early Bf 109 F-1 (prior to external strapping) was indistinguishable from the late Bf 109 F-2!

Some months later, the Bf 109 F-3 appeared. Cannon armament reverted to the old MG FF, presumably because of limited availability of the newer Mauser weapon. There is little to aid distinction of this aircraft from its predecessor. I can offer but one clue! Replacing the Daimler-Benz DB 601 N engine was the more powerful DB 601 E. This motor used 87 Octane fuel, as indicated by the figure '87' stencilled on a yellow triangle close to fuel tank filler panel on the left fuselage wall, below the cockpit canopy. Both the F-1 and 2 used 96 Octane, which was called 'C3' and this figure was shown on the triangle.

It is apparent from our research that very few Bf 109 F-3 aircraft were built, and this is supported by the almost total absence of reports of its use operationally. The MG FF installation was troublesome, and a failure of this weapon rendered a fairly lightweight 'punch' almost impotent. It seems likely that production of the F-3 was terminated in favour of the similarly-engined Bf 109 F-4, by far the most numerous of the series, in which the central armament reverted to the MG 151/20. In a belated effort to improve the firepower, this sub-series was equipped to carry a further MG 151 in a gondola under each wing. It was the only F-series with this capability. Lastly, and perhaps not as evident, armour protection for the pilot's head was enhanced with the addition of a curved panel to the top of the near vertical plate which was standard on the F-3.

Two further marks were produced in minute numbers, both tactical-reconnaisance machines. The Bf 109 F-5 and 6 had a camera (typically an Rb 30/50) inserted behind the cockpit and ahead of the radio racking. Pointing straight down, its presence is detected by a shallow, streamlined fairing under the fuselage, aft of the wing trailing edges, protecting the camera lens from rain and escaping engine fluids. Detailed information is scarce, but it is believed that in both versions, the engine-mounted cannon was deleted. One source claims the F-6 to have been devoid of all armament. Personally, I find it hard to believe that there was any need for, or logic in, an aircraft in the low-level reconnaissance role being left completely defenceless. It is probable that both, rather than being new-build, were produced using Bf 109 F-3 and 4 airframes.

Prior to its introduction to service in the North African desert, modifications were made to suit the environment. The supercharger intake was made more robust in order that a *Sandabscheider* (sand filter) could be bolted to it. This is easily distinguished from the slimline version by its rounded lip into which were inserted eight anchor nuts to attach the filter. Fitted to the Bf 109 F-2 for this purpose, it became standard equipment on the F-4 and is a useful pointer. Further, in the F-4 a larger oil cooler was introduced to cater for the higher ambient temperatures of the desert. This resulted in a deeper intake under the nose and larger 'bath'. Another sign of tropicalization was the use of whitewall tyres, but these were by no means standard fit in service. So configured, the aircraft became the Bf 109 F-4/Trop and earlier aircraft, although not so fully fitted (they did not have the deeper oil cooler, for example) were designated Bf 109 F-2/Trop.

At the risk of confusing the reader further, a *Werkschrift* indicates that it was possible to exchange an MG 151/15 for a /20, and vice versa. Thus, a Bf 109 F-2 could be fitted in the field with the large weapon and the F-4 with the smaller!

I did say that identification of one 'F' sub-series from another is difficult! It would be more accurate to state that in a great number of photographs it is impossible because of the absence of detail. Barring background information gleaned from operational records, I think it is fair to comment that identity is, at best, a guess.

Distinguishing the Bf 109 F from the Bf 109 G, and the early *Gustavs* from each other, conversely, is simplicity itself. Built in vast numbers, the Bf 109 G

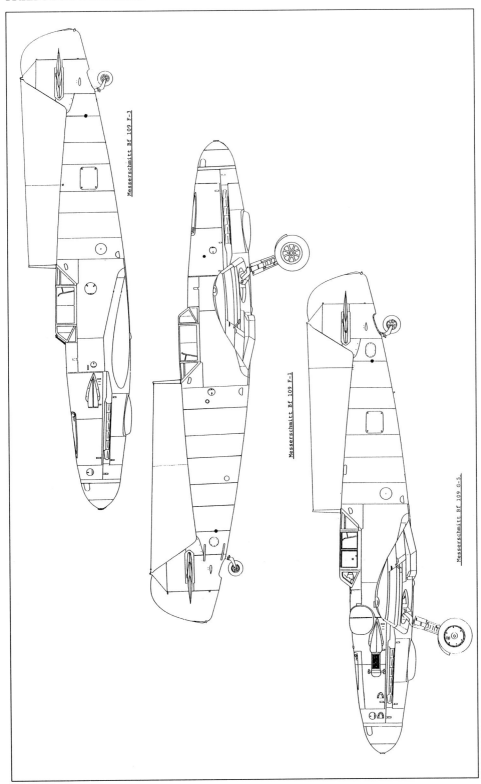

Messerschmitt Bf 109 F-3

Messerschmitt Bf 109 F-1

Messerschmitt Bf 109 G-5

incorporated the more powerful DB 605 engine and a selection of improvements to the *Friedrich* airframe. However, as there was a delay in the new engine programme, twelve pre-production Bf 109 G-0 were produced powered by the DB 601 E. Externally, the only significant difference in appearance — and the most important feature by which the *Gustav* may be recognized — lay in the new windshield and cockpit canopy. The former, with armour glass in-built, bore very heavy framing compared with the earlier assembly which had a flimsy appearance, and the canopy itself comprised heavier framing.

Production machines first appeared in the summer of 1942 and displayed a number of new features, these serving to distinguish one mark from the other and the early marks from the Bf 109 F. Firstly, coupled to the DB 605 engine was a VDM propeller with blades of greater chord (wider appearance) to absorb the greater available power. The upper engine cowlings retained the bulkier supercharger intake which was introduced on the Bf 109 F-4, and, therefore, all 'G' cowlings could be fitted with the sand filter already mentioned. The lower cowling also retained the larger oil cooler in its deepened housing. However, as there were instances of overheating during testing, a small intake was cut into the cowling immediately behind the propeller which encased the upper half of the annular oil tank. Immediately aft of this, an identical scoop was fitted to the front end of the upper main cowling.

Further back, the fuel tank filler point, formerly on the left cockpit wall, moved to the spine of the fuselage on the *Gustav* and its panel can be spotted on the left side, placed between frames 3 and 4. Towards the tail unit, a large oval inspection panel (to improve access to the tiny rear fuselage and its equipment) made an appearance on the left side, between frames 8 and 9. In the same region, the tail undercarriage leg became non-retractable and the recess into which it formerly retracted on the 'F' gained a solid magnesium alloy frame onto which could be screwed a fairing which served to exclude debris thrown up by the wheel. I say *could* be screwed as it was fairly common in practice for this to be absent; but the exposed framing is generally obvious.

The mainplanes were almost identical in appearance to their predecessors. The inboard flap arrangement for the coolant radiators was altered, but this is only evident in close views. The wing tip navigation lights, formerly mounted in a scalloping of the edge, were enclosed in a plexiglass fairing. Seen from below, a significant new feature was a mainwheel well with a straight outboard perimeter; that of the F-series was circular.

These, then, are the features which point to an aircraft being a Bf 109 G rather than an 'F'. Now to features which distinguish one 'G' from another! With the introduction of the type, provision was made for a pressurized cockpit to make life tolerable for pilots facing combat at high altitude. Pressurized and non-pressurized versions were produced simultaneously. The former bore odd mark numbers, e.g. Bf 109 G-1, 3 and 5. Thus the first two versions are easily recognized. The Bf 109 G-1 bears all the features described in the preceding paragraph. The Bf 109 G-2, on the other hand, carried a small air scoop below each windshield, plus an inward-moving rectangular ventilation door on each cockpit wall, both these pointing to a non-pressurized cockpit.

The G-3 and G-4 closely resembled the the G-1 and G-2 respectively. However, the radio fit changed from the 1935-vintage FuG 7a Telefunken design to the FuG 16 and moved closer to the tail. Therefore, the lead from the antenna can be seen entering the fuselage at a new position just forward of the fin. More easily spotted, the newer aircraft carried new wheels and larger tyres. The cast, spoked mainwheels and tyres (size 650 x 150mm) were replaced by pressed wheels of smooth appearance clad with broader covers (660 x 160mm). The mainplanes were also altered to accommodate these larger tyres, a panel on each upper wing surface sporting a pressed, semi-circular bulge over the outer wheel well. The tailwheel, also larger, was much bulkier in appearance.

A matter of a few months after the introduction into service of the *Gustav*, the most important mark became available. This was the Bf 109 G-6 and, to a lesser extent, the Bf 109 G-5. The G-6 was eventually built in bewildering variety, and even dedicated researchers have come to grief attempting to catalogue them! Recognition of the early production models can be readily accomplished. In effect, the G-5 and 6 were G-3s and 4s with improved armament. In place of the two rifle-calibre MG 17 machine guns ahead of the windshield, they carried two MG 131 guns of 13mm calibre. These more bulky weapons necessitated the modification of the upper cowlings and a large, near-circular fairing protruded at each rear edge, marring the smooth lines of earlier panels.

Like its forebears, the *Gustav* could be modified to suit field conditions. The 'tropicalization' adaptations are clearly listed in the *Ersatzteilliste*. Internally, a survival pack was installed in the luggage compartment behind the pilot's head. In the rear fuselage, a Mauser K98 rifle was strapped to the starboard wall. Leathercloth sleeves were fitted over hydraulic actuator piston rods to exclude sand and dirt, and whitewall tyres were supposed to be fitted, although it appears to have been more common that only the tailwheel was so clad. Of course, the filter was fitted to the supercharger intake but, interestingly, another listed 'Trop' modification (on non-pressurized aircraft) was the rectangular ventilation intake on each cockpit wall. As I have yet to find a picture of a non-pressurized Bf 109 G *without* these, I can only assume, like the fitments for the supercharger filter, they were adopted as standard equipment. Absurd though it may seem, the only indication that an aircraft has been fully 'Trop'd' was the appearance of two small, streamlined blisters on the port cockpit wall, below the windshield framing, one above the other. Their purpose was to hold a large parasol, which would have afforded some shade to the small cockpit otherwise exposed to the desert sun.

Most of the information I have laid out is derived from the documentation gathered during the project. All of it has been available to researchers of the Messerschmitt, yet some myths continue to be perpetrated. Two, I feel, should be finally laid to rest! First, it is often claimed that the Bf 109 G-1 was armed with the MG 131 on the upper decking. Not so! It was identically armed to the Bf 109 G-2, 3 and 4. Second, far from being a reconnaisance fighter, as described by many authors, the Bf 109 G-2 was a standard front-line fighter.

6

Work in a shed

Our ejection from a well-appointed hangar, at very short notice and for no apparent reason, rankled for some time, especially as I could see our former workplace lying empty. However, as I had already lost months of valuable time in the move from Lyneham, I was anxious that the project should resume as quickly as possible, and I suppressed my indignation.

Our new home was a large building roofed by sheets of corrugated iron, and used for years as a depository for redundant equipment and materials. It was rarely, if ever, cleaned out. With the total absence of windows, a shaft of daylight from the open door penetrated the gloom like a searchlight. Deeper into the shed, it was suitable only for finding one's way about, and was supplemented by wan lighting of wartime origin high in the arched ceiling, two units of which were inoperative. Even in the summer the atmosphere was humid. There was no heating or proper ventilation, and worse, it was evident that the roof leaked copiously. Barring power for the minimal lighting, there was no electricity supply. As a potential workplace, Building 120 was a Dickensian nightmare! I immediately informed my superiors that it was unacceptable in every respect, and I received placatory assurances that improvements would be made.

That hurdle apparently overcome, a lot of hard toil lay ahead before work could resume. Although the 109 lay just inside the shed, the allotted workspace was at the other end! But it was far from being a simple case of moving everything thirty yards, as the selected corner was buried under a weighty heap of assorted ironmongery and a substantial pile of soggy sand! The former was little better than scrap, whilst the latter bore striking witness to the inadequacy of the antique corrugated canopy, both inexplicably occupying storage space considered to be at a premium at Northolt. There was nothing for it but to set about lifting and shovelling. The cleaning process was completed within a few days, albeit with the most welcome help of some civilian staff from the station workshops. The concrete floor, once exposed, proved to be in poor condition, the decay doubtless accelerated by the wet mass it once supported. The last thing I wanted was the aircraft covered in a film of concrete dust, so I carefully swept the floor to remove all the loose debris and then applied a coat of sealant. A few days later, the Team inserted everything into a relatively tidy corner, and work resumed.

Whilst Ian and I completed the stripdown and clean-up of the fuselage, Tony Leek tackled the tailcone with some assistance from a new arrival, Kevin Thomas. Also receiving intermittent attention at the time was the cockpit windshield assembly. Fairly crudely assembled, the rough steel framework was easily stripped and de-corroded, after which a couple of coats of grey-green primer were expertly applied by Don Silver, one of the extraordinarily helpful staff from station workshops.

I have been criticized for not coating the airframe in a modern material, such as etch primer, which would have offered more protection. Whilst I accept the argument, it must be borne in mind that my aim was to *restore* the aircraft, and it was my intention that it be finished correctly, even to the extent of the primer. Virtually the only original paint surviving on the aircraft, this was found on large internal areas of the fuselage, particularly in the fuel tank bay. Here and there, loose flakes could be found and these proved a perfect match with paint samples illustrated in the Luftwaffe Manual of Painting, L.Dv.521. (In later years, I found evidence of many variations in the shade of such primer, in Luftwaffe terms, RLM 02. An example was that found inside a Heinkel He 111 P bomber which was beautifully rebuilt at Oslo Gardermoen. This was much darker and more akin to paint used by British industry.) Armed with this evidence, I approached Berger paints in Bristol to seek their help. Remarkably, on their shelves lay three large cans of grey-green zinc chromate primer; old stock which was scheduled for disposal. Although close to the Luftwaffe shade, it was not quite right, and Berger kindly instructed their laboratory staff to tint the remaining stock prior to presenting it to me. It was a very dense substance, and required a great deal

A nice clean Bf 109 G. The hard work had yet to start!

With the fuselage clean complete, Ian Mason and author have a brief rest.

of thinning prior to application. We were confident, therefore, that we had more than enough to complete the aircraft.

The basic fuselage 'clean and repair' was completed in late January and, during the course of one marvellous afternoon, it received a coat of primer from Don's gun. The result was beautiful to our eyes and provided a much-needed morale boost. It was also a complete vindication (if one was needed) of my decision to perform a thorough restoration.

Shortly after, the tailcone received the same treatment and we bolted it to the fuselage. Casually said, but it took many hours to fasten the rows of nuts

Left *View of front of fuselage showing the all-important 'D'-frames at the bottom, engine bearer lugs at top, and wing attachment points to the side.* **Right** *Inside a spotless rear fuselage. The restricted access towards the rear is most obvious. Towards the top, the circular hole to the left reveals stowage position of Mauser survival rifle.*

with new split pins — another job which entailed tongues probing thin air! However, the reconstruction had begun! We forgot our primitive surroundings temporarily, and countless parts, large and small, received detailed attention. I devoted weeks of work to the tailplane, which proved extremely difficult to clean due to the very limited access, yet it took mere minutes to mate it to the fuselage. The fin was a much easier proposition and, with that fitted, the rear of the aircraft began to assume the unmistakable form of the Bf 109 for the first time in five years.

During this phase I arrived for work one day to find Ian and Kevin Thomas beavering away. With them was a young man who was introduced to me as John Elcome. John had sought the permission of the Station Commander to photograph the Spitfire XVI which served as 'gate-guard' at Northolt. As it happened, this hulk was in Ian's charge. He spotted John examining it one day and, unaware that permission had been granted, sauntered over and introduced himself with the question: 'What do you think you are doing, then?' — or words to that effect! The situation quickly resolved, Ian invited him to have a look at an interesting aircraft! From this accidental but fortuitous meeting the team gained another member. A British Telecom engineer by trade, John was unqualified in aircraft engineering, but he threw himself into the most difficult, and generally tedious tasks with genuine dedication. I don't ever recall him complaining, and his work was first class. He was also a photography fiend, and his activity over the ensuing years in recording our progress is evident in the pages of this book.

Whilst we laboured, winter set in with a vengeance. When the wind blew, we were appalled to see the roof panels above our heads lifting and separating. Fortunately, they held, and we became used to the alarming clattering all around us, amplified by the cave-like qualities of the shed. As the temperature inside neared freezing point, we resorted to stuffing reams of old newspaper in every crevice likely to allow the bitter breeze to penetrate. Thankfully, the overhead bulb/shade combination was replaced with neon lighting and, at last, we could work with the doors closed! We were forced to take drastic steps when electrical power was needed. The sole junction box lay over a hundred feet away from the aircraft. We attached a cable to suitable terminals within and draped it the length of the shed. Inevitably, there was a drastic loss of power at the other end, and the cable became extremely hot at times, but at least we could use a power tool, boil some water, or use a small fan heater. I was very worried about the effect of the very damp atmosphere on the exposed fittings, and we regularly positioned the heater beneath the fuselage, directing the warm air into the fuel tank bay, from where it wafted into the rest of the assembly. However, it became so bitterly cold on most days that we took turns heating our numb extremities instead!

Of course, I had broached the subject of heating with the Station Commander. He, in turn, discussed the matter with the local manager of the 'Ministry of Works and Bricks'. The conclusion had been entirely predictable; heating the shed was out of the question on the grounds of cost. But this I had anticipated, and suggested instead the erection of a partition parallel to the end wall to form a small bay around the 109, the exposed front of which could have been draped with a suitable material and the enclosed space heated at a fraction of the rejected cost. The plan was accepted, and an amazing screen of heavy timber framing took shape, perilously close to our vulnerable aircraft, to which was pinned fibreboard. On the shed wall, electricians installed conduit and cabling suitable for heaters, and then ... nothing! After a few weeks of puzzling inactivity, I learned that there was no cash left with which to complete the work! Nothing further was ever done towards completing the proposed improvements.

The cockpit received a coat of RLM 66, a dark grey colour, which gave the area an 'as new' appearance and beautifully highlighted the various metal labels. The next step was my responsibility solely. On the left wall, three words denoting the action of the tailwheel locking mechanism had to be reproduced freehand in white paint. I spent some hours practising until I was sure, before entering the cockpit, that I could form the letters in the style of the Messerschmitt employee. As may be imagined, the cramped conditions made a good posture impossible. Coupled with the knowledge that one slip would be disaster. I had good reason to feel apprehensive as I applied the first brush stroke! It all went well, but I was very relieved when I finally put the brush down.

On the opposite wall, we had discovered the word *Netzeinschaltung* — battery switch — and an arrow. This, however, was in the form of a decal. I contacted a specialist, to be told that the cost of producing one or two such decals by prototype methods would be prohibitive. But he agreed to examine my requirements. Some weeks later, I was delighted to receive two superb decals from Eagle Transfers of Lichfield, free-of-charge, and one of these I applied to the aircraft. Small though the item was, it took considerable effort and time on their part to produce, and the gesture was much appreciated. In contrast, some years later another decal company (Eagle having foundered, regrettably) produced me some decals for engine-mounted tanks. The process took months, during which time my instructions for the order were lost at least twice, and the first effort was twice the required size! The correct size was eventually produced, but with a spelling error and, amid much complaint, a third attempt — finally acceptable — arrived with a substantial invoice attached!

We then began installing systems in the fuselage, starting with the stabilizer trim which comprises a screw jack connected to a large handwheel in the cockpit by 'bicycle chains' and cables. The woodwork of the wheel was restored by workshops, new plywood being fashioned to match the original. Much to our delight, its operation matched the manual description and no adjustment was needed.

My letters seeking help began to bear fruit. So many had I typed that a format developed which required only slight amendment to suit each appeal. In fact, I had become an expert in begging letters! Many of the control tubes for the flying controls had been bent or severed, and several were devoid of their fork-end fittings. In response to one of my missives, British Aerospace Training Division at Hatfield offered help, and Ken Wheeler visited Northolt to discuss the problem. He left, not only with a selection of tubes to repair, but also two main undercarriage legs!

As we were fast approaching the time when hydraulic components would be re-fitted, I contacted Dowty Seals of Tewkesbury, and replacement seals were kindly offered following laboratory spectrophometer examination of the remains of some originals. Meanwhile, through Peter Nolte, contact was re-established with Drägerwerk of Lubeck. I had acquired an original oxygen panel in seemingly excellent condition and delivered it to Peter who submitted it in turn to Dräger, the original manufacturers. The result of their examination was disappointing; they could not return the regulator to operational health. However, instead they offered a modern substitute which was a direct derivative of the 1942 equipment. This I gladly accepted and arranged delivery of a completely new oxygen system, again free-of-charge. It was to be some time before I was in a position to install it.

Various other pieces of equipment, however, were refitted after a complete stripdown. The throttle quadrant was carefully cleaned, but I retained as much evidence of the original markings as possible, following removal of the Meteor fuel cock and repair of the damage. Everything repainted, with markings for throttle position and fuel cocks restored, it took its place on the 'pillar' of the left cockpit wall. The connecting linkages were also cleaned and found to be in sparkling condition. Once again, we had evidence of the value of anodization! Interestingly, one rod bore the stamp 'He 177'. The Heinkel He 177 was a complex four-engined bomber. I can only surmise the control rod to have been a replacement after 10639 found its way to England.

In fact, words of one sort or another were found all over the aircraft. Mainly these were inspection stamps whose number and variety were bewildering! The more critical the component, the greater the number of stamps. Delicate equipment had received a dye mark, while the more robust had the relevant details punched into their surfaces. As the cleaning of the aircraft progressed, we anticipated the discovery of such evidence of inspection. Where none was found, the authenticity of the part was suspected, and with very few exceptions our suspicions proved justified. Furthermore, all major structures carried identifying alloy labels and these too had inspection marks upon them, presumably indicating that the completed assembly met the requisite standard.

It was a great pleasure to all the team to be involved in reassembly. Month after month of tedious cleaning had eroded the spirit. Even with the greatest degree of motivation, such work, in near-freezing, primitive conditions is not recommended! The satisfaction we all felt in putting the first parts together was all too evident by the level of banter during these work sessions.

7

Progress

The fuselage, with its windshield and tail surfaces in position, looked exactly like an aircraft on the wartime production line. I indulged in a brief period of self-congratulation. We were the recipients of much complimentary comment — a novel experience for all concerned. As we had been effectively hidden in a far corner of Northolt, this may seem surprising. (Our isolation was most welcome as it allowed us to work uninterrupted.) However, news of our progress suddenly put us on the compulsory viewing list for visitors! The Station played host to many officers of Air rank, on official duty more often than not. The casual visit caused us little pain; we became accustomed to high-ranking officers walking into the shed in the company of the Station Commander, and it was usually a pleasure to have a conversation with them.

The official visits were a different kettle of fish. Weeks before the event I would receive orders to be present in the shed, normally with Ian Mason. We had to devote hours cleaning the area and making it tidy. We were not normally untidy, you understand! However, the shed was inherently dirty and, without the screening I had demanded, a thick layer of dust settled on all our equipment in the blink of an eye. For this reason, everything had to be covered. Not all visitors were knowledgeable in Second World War aircraft, and one such, an Air Marshal, having surveyed our work for some minutes, was heard to say to the Station Commander: 'Its lovely ... but what is it?'! I suspected that, in arranging these occasions, my boss hoped to provide some light relief for his guests. An uncharitable thought, perhaps, but it did strike me that there may also have been a diversionary motive, distracting attention from another part of the camp!

One distinguished visitor brought all work to a standstill for a complete afternoon, yet not one minute of that time did we begrudge him. He was Wing Commander Robert Henry Gibbes DSO, DFC and Bar, Royal Australian Air Force (Retired). I had discovered that Bobby — or rather his Squadron — had been the captor of 10639 (of which more later). One of

Ian Mason shows Bobby Gibbes an aeroplane he had not seen for 40 years. He was not allowed to take it home!

the foremost Aussie Aces, he scored ten confirmed victories while flying his Curtiss P-40 Kittyhawk in the desert, added to which were 14 probable 'kills'. Post-war he turned his hand to operating a fleet of Austers in New Guinea. These were later supplanted by the larger Norseman and even larger Junkers Ju 52 transport aircraft. Subsequently he settled in Sydney. I was delighted when a friend of his, Air Marshal 'Bing' Cross, asked if Bobby could see the Bf 109 while on one of his periodic visits to Europe. I will always remember his words as he spied the old aeroplane. 'It's just like new!' Suffice it to say, he was overwhelmed to see his old 'prize' in good health and receiving a generous dose of tender, loving care, but I had to refuse his one request — that I return the aeroplane to him!

Between visits, several sessions each week were devoted to the rebuild of the fuselage, but eventually we ground to a halt in almost every direction because of a lack of parts. I continued my search for replacements, but appeals placed in the aviation press availed nothing. Amazingly, only once did my efforts bear fruit over the many years of the restoration. I am convinced that much potentially useful German equipment lies in attics gathering dust having been brought home as souvenirs. Simple parts, of course, could be reproduced, and these I placed with craftsmen at Northolt and with various companies. The staff of the RAF Museum Unit at Cardington offered help where they could. Chief Technician John Wadham volunteered his fabric skills by re-covering the five flying control surfaces, following German methods, but this work bore no priority and the Museum was never at a loss in finding an aircraft for him to finish. Consequently, it was to be a couple of years before Messerschmitt parts received his ministrations.

Reconstruction has started. Tail and front bulkhead refitted and the fuselage coated in primer.

Missing components, for example, instruments, switches, electric plugs — even aviation-standard metric nuts and bolts — had to be found and acquired. The only quick way around this obstacle would have been to fit substitutes, but this would have made a mockery of the restoration. Where

Left *Pleasing view of 10639 with canopy temporarily in place.* **Right** *Windshield in place and front bulkhead adorned with some control linkages and hydraulics.*

equipment was placed with individuals or companies for refurbishment, it was accepted on a voluntary basis. There being no money involved I could hardly insist on speedy delivery! Thus, I had difficulties with anything outside my direct control.

Exasperating though it was to leave some work unfinished, much further work lay in the wings — literally! The whole team turned its attention to the port mainplane which had lain, totally neglected, since our incarceration in Building 120. Correction, it had received close scrutiny during the course of one day! In fact both wings and the tailplane had their spars and critical fittings examined using Non-Destructive Testing techniques (NDT) by a team of experts from Brize Norton. They also checked the 'D-frames' on the fuselage — particularly important for reasons already explained — and the centre spar, plus all mainplane attachments. All passed with flying colours. There were no defects. Whilst we were at Lyneham, a sample of fuselage alloy skin was submitted to Harefield laboratories for age-hardening tests. We supplied the worst bit of metal found, yet it, too, was deemed healthy!

Each man, armed with a brush and cup of synthetic thinners, selected an area of wing and set about the task. Seeing them working together on one part of the 109, buoyed with enthusiasm as a result of recent progress, I was very conscious that, had they not volunteered to help, the project would have been abandoned. I could only marvel at my good fortune. As most areas of the wing are easily accessible, it took very little time to complete the bulk of the cleaning and remove traces of surface corrosion. Later on, John Elcome, Kevin Thomas and myself tackled the enclosed areas. By far the worst of these was by the main spar, near the root, and the ammunition compartment for the MG 151 cannon. Access was terrible but, then, they were not designed to facilitate cleaning by the British 40 years later! Some

The starboard mainplane before internal strip had commenced.

The restored throttle quadrant. Small lever is the fuel cock.

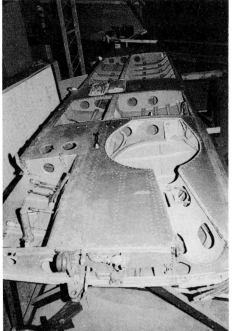

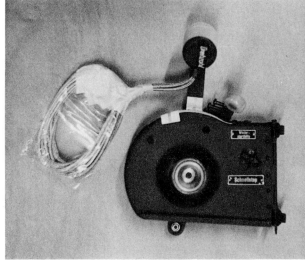

The simple construction of the wing. Its accessibility is apparent.

spots could only be reached by putting an arm though one hole and working by touch, or torchlight and a mirror seen through another. We each had difficult parts to tackle, but only Kevin, with his small hands and slim (skinny, to the rest of us!) wrists could hope to reach the worst. To do so, he was forced to adopt the most ungainly positions!

I remember him draped over the upturned wing, one arm totally immersed in the structure, struggling to reach the furthermost corner, and tongue out to full extension, moving in sympathy with his hidden hand as if attempting to lick clean an invisible part. Making rude remarks about his posture only prolonged his agony. His tongue instantly retracted to allow a smile to develop, and it was a fair bet that his hand had stopped work! Those confined regions proved extremely difficult and we were kept busy for some weeks, occasionally exchanging duties, if only to peer into a new crevice. Not infrequently, a sharp edge or rivet tail caught a probing finger, or perhaps knuckle. The ensuing oath, far from eliciting sympathy, was generally followed by a rejoinder from Ian not to 'bleed on his aeroplane'.

The oil cooler and two coolant radiators had been in the care of Delaney Galley, a few miles from Northolt. After a respectable time I telephoned to learn the result of the examination. The news was mixed. The brass oil cooler was thought to be serviceable, but a full test proved impossible because a pressure-relief valve was missing. The radiators, being matrices of aluminium alloy, were beyond repair. Whilst this was disappointing, the situation was not yet desperate. Included in the exchange deal with the Finns were two further radiators. However, the history of these was unknown, and it was likely that they, too, had internal corrosion damage. Therefore, they received a deal of close scrutiny upon arrival. Both appeared reasonably healthy, but not exactly pristine, and I approached Delaney once again. This time the reply indicated a service charge which was impossible for me to contemplate. Next, I sent one of my begging letters to Serck, whose reputation in the field of radiators was second to none. The telephone rang days later and I was able to discuss the situation in detail. The first suggestion was a simple investigation to be carried out in the shed — stand the radiators on end, fill them with white spirit ... and watch! This we did with all four, the original two instantly revealing the extent of their deterioration, with fluid escaping faster than it could be inserted. Indeed, they gave passable impressions of ornamental fountains!

The Finnish pair gave more cause for optimism. Twenty-four hours later, three spots of dampness were evident on one, while the other remained completely dry. Armed with that information, Serck approached their sister company, Behr, which subsequently agreed to an overhaul, provided the radiators could be delivered to Stuttgart. Some weeks later, British Airways flew them from Heathrow, free-of-charge, this again as a result of my letter writing. Behr, incidentally, was the original manufacturer. I had no idea that it was still in business!

By a stroke of luck, I wrote to the Swiss Air Force Museum, hoping to find Messerschmitt parts in their store, just at the time when the Pilatus P-2 training aircraft was about to be withdrawn from service. This venerable machine had been designed, for cost reasons, to incorporate equipment from the Bf 109 E which the Swiss formerly operated. Its undercarriage, for example, taken from the 109, was adapted to allow the main legs to retract inwards, affording the trainer a much broader track. The Swiss offered all the Messerschmitt equipment, plus other parts, including a Revi C/12D gunsight, but in exchange for a rare engine! I foresaw little hope of being able to close such a deal, but approached the indispensable Jack Bruce. Against all the odds Hendon had such an engine in store — a spare which had been sectioned for display purposes — and it could be made available to me. Thrilled by the news at first, I later wondered whether the parts offered by Dubendorf were worth the sacrifice of a huge engine. It transpired that, in a previous exchange, Hendon had gained a De Havilland Venom fighter in return for very little and there had always been an intention to properly repay the Swiss. In helping the 109, therefore, a debt would be settled. Once again, the greatest expense incurred was crating the engine.

Christmas came early that year when two crates arrived at Northolt. Crammed with equipment, the undercarriage legs particularly drew our attention. Although modified by Pilatus, they were to prove very useful later. The tail-leg could have been fitted to the 109 within minutes! Close by we found two beautiful mainwheels and a tailwheel. We badly needed all three, as our mains were in poor health and the Spitfire tailwheel was of even less use. Added to these were two wheel-brake units which were as new, having been totally refurbished by the Swiss. The only disappointment was the non-appearance of the promised gunsight, but we learned that it was to be sent separately. (It never did reach us as it was intercepted by staff at Hendon who unwittingly fitted it to their Bf 109 E!) Nevertheless, it was a real windfall for the project.

Progress at this stage was most encouraging, problems notwithstanding, and it was steadily maintained despite the end of my Commission in the Royal Air Force. I had set up home in Brackley, Northamptonshire, and begun a second career in the airlines. The journey to work on the 109 ceased to be a few minutes walk and became a drive of one hour and a half! I scheduled a weekly session with Ian and occasionally managed two. As much work was performed by the team when I was not in attendance, my worry was not so much our progress, but rather the future of the bulk of my work-force still subject to the vagaries of the military posting system, and I began to contemplate contingency plans. I was pleasantly surprised, therefore, to hear from John Danes, a Service Manager of Rolls-Royce in whose charge the engine had been. He suggested that the project was moved to Filton, and he had already been granted hangar space and found many willing vol-unteers. It seemed the ideal solution, should my worst fears be realized.

8

Unwanted attention

Not nearly so welcome were the earlier attentions of a junior engineering officer at RAF Coningsby. The phone rang and I found myself relating the details of the project to an apparently sympathetic listener. After some minutes alarm bells sounded and I asked the reason for his call, connected as he was to the Battle of Britain Memorial Flight (BBMF). His answer was abrupt and astonishing in the arrogance it displayed. 'Well, I doubt whether you are capable of completing the project. We have all the expertise needed here and I want to take over the 109. Who do I speak to about it?'!

I feel that there is no need here to describe my answer! (I wondered then whether this brash individual knew anything at all about the aircraft he proposed hijacking. Somehow I doubt it.) I had no need to worry that he, personally, might succeed in his takeover bid, but there was the strong possibility that he was probing on behalf of his masters. I quickly phoned the overburdened Jack Bruce at Hendon with the news, and asked him to be ready to support me and my efforts. Just in time, as it turned out! Jack, 'without even waiting for the telephone to cool down' (a letter later revealed) called Wing Commander O'Reilly in Whitehall in whose charge the Bf 109 and all the MoD-owned historic aircraft lay. Mid-conversation the good Wing Commander was dragged to another line on the end of which was a Squadron Leader from Coningsby! Fortunately, Whitehall retained confidence in my work and, although the unidentified officer argued well, he was told to back off. I was pretty disgusted by this turn of events, especially as it was quite likely that BBMF would eventually inherit the 109 in any case. I registered my anger with 'Jacko' Jackson, the Officer Commanding BBMF (and formerly the Qualified Flying Instructor on Varsity aircraft when I was stationed at RAF Gaydon) in the hope he would realize that, rather than expedite the restoration, BBMF interference could very well have placed it in jeopardy. His reply indicated that there would be no further intervention, but that 'a closer interest will be kept on your aircraft in future at a very high level'. How very true this warning proved!

Equally annoying and distracting, the new engineering officer at Northolt showed signs of being a nuisance. Late in 1981 he ordered Ian Mason to produce a detailed situation report together with his projections for the future of the project. Using this, he transformed the last paragraph by revealing that 'Northolt in conjunction with ... Field Aircraft Services was examining the possibility of providing more formal support.' Such assistance would have been welcome had it been openly offered and in an acceptable form. However, as I learned of it only after a copy had been delivered to Hendon in the form of an appeal, not unnaturally I took a dim view of this underhand move.

Around the same time, Ian reported that an Air Vice-Marshal had been shown the aircraft by the same engineering officer. Enquiry revealed him to have been the Air Officer Commanding 11 Group under whose wing lay ... BBMF! I was unable to extract the reason for his visit, but later, through some astute detective work by Ian, I discovered that one of his staff officers had approached the Historic Aircraft Committee Chairman, his brief being to investigate the possibility of transferring the aircraft to BBMF! This approach was also rebuffed.

Why BBMF and 11 Group chose to show such unwelcome interest at that particular time is not known to me, but as it coincided with the appeal issued by the engineer at Northolt, there is an obvious connection. In proposing to move the 109 to Coningsby, the 11 Group staff officer promoted the engineering facilities there, although, in the same letter, he admitted that BBMF had 'no spare capacity'. (Those same facilites existed at Northolt, of course, but I had no official sanction to use them.) I have no doubt, with suitable finance — substantial finance — BBMF could have rebuilt the aircraft, but only after a great deal of research. If it was ever intended that it be accurately restored (and I detected no interest in this direction), even more research would have been required. In summary, therefore, I fail to see how progress could have been accelerated with the aircraft in BBMF hands. The most appalling aspect of this whole sorry episode, to my mind, was that neither the Northolt officer concerned, nor AOC 11 Group ever saw fit to talk to me, the person they both knew to be in charge of the aeroplane. I cannot forgive the secretive way in which all their activities were conducted, and, even ten years later, the anxiety and worry they caused me is fresh in my memory.

1982 was not a memorable year, all in all. We lost Pete Hayward when he left for 'civvy street' and, shortly after, Tony Leek was posted to St Athan. Bad enough, but more followed! In April, during a journey to Wolverhampton to deliver our fuel system to IMI Marston, Ian and I learned of the invasion of the Falkland Islands by Argentina. We both knew instantly that the reverberations would affect us badly, and it was not long before we discovered just how badly. All the work that had been placed with Abingdon and other stations was put on indefinite hold while the Service was placed on a war footing.

The Team now numbered four and, for the first time, I worried whether I would be able to continue in such reduced circumstances. We persevered for some months finishing the left wing which, like the fuselage before it, looked magnificent in its coat of grey-green primer. The extensively damaged plain flap was substantially rebuilt and the radiator flaps repaired.

My nightmare became reality during September when Ian learned that he was to be sent on a six-month detachment to Ascension Island in the South Atlantic. Lonely Ascension had become supremely important to the RAF as the staging post for aircraft supporting the regained Falklands, and a vast amount of manpower was required to man the airfield and its services. General Galtieri had a lot to answer for! The only glimmer of hope for me was that, after his exile, Ian was assured that he would resume his position at Northolt.

Thus, over winter, Kevin Thomas, John Elcome and I did what we could at weekends, but this was really confined to cleaning work. We turned our attention to the last major component, the starboard wing. This was familiar territory, and we all knew the problem areas to be tackled. But, strangely, this wing was finished differently from the other internally; it was entirely sprayed, whereas the left wing only had primer crudely applied by brush in an effort to allay the onset of corrosion on areas where two dissimilar metals met — for example, where a bolt passed through an alloy rib. As a result, it proved to be in excellent condition. Marks made by the German constructor were found on ribs, providing an interesting diversion from the tedium of cleaning as we attempted to understand their meaning. We failed, I might add. The scribble was worse than mine on a bad day!

Some months previously I had been warned that plans were afoot to redevelop our building, and that it was possible that I would have to move the aeroplane — again! This time, though, no alternative accommodation was offered. The Messerschmitt was to be ejected from Northolt. However, the Falklands War side-tracked any such plan for many months. I suspect that the defence budget had been devoted to more important matters. But the threat was never far away.

The new engineering officer made his views known. 'I would prefer to see the removal of the whole project from 120 Hangar and Northolt, so that the hangar may be put to good use for storage of snow and ice clearance equipment and other good service uses.' It was clear that little sympathy, let alone assistance, would come from that quarter. Weeks later, by dint of reorganizing in the building, the snow clearing clobber was easily accommodated, but I had not seen the last of the opposition to our presence.

Ian returned to the fold in March, promoted to the rank of Chief Technician during his sojourn in Ascension. Good news though this was, a posting attached to the promotion made the outlook very bleak indeed. He was directed to start a new job at Gütersloh in Germany with 230 Squadron in mid-July. It was an interesting posting and I was not inclined to interfere. Thus, we had three months of Mason, the team restored to four — but only temporarily. Kevin Thomas found himself on the road to Odiham shortly after!

There could be no question of the project continuing with John Elcome and I as total workforce. The pressure which was being steadily applied by the engineering squadron at Northolt, through the office of the new station commander, to vacate the premises was, in any case, intolerable, and I decided to start afresh in a new home. The crunch came at the end of June. Summoned before the Group Captain, I was given until the end of the following month to pack up and leave. This time the reason given was the proposed development of our shed as an engine overhaul facility! Both he and his engineering officer, who was in attendance, blocked any counter proposals I made to prolong my stay. I would like to think that I am an even-tempered individual, but I stormed out of the office seething with rage and indignation.

I have mentioned that John Danes at Filton had proposed moving the project there. Unfortunately, the plan foundered in tragic circumstances. John entered hospital for a serious operation and then began a period of convalescence. He eventually resumed a light work schedule, but died soon after. We had lost an avid supporter of the project and the mainstay of the engine rebuild. It transpired that John had negotiated hangar space on a personal level but had not been able to formalize arrangements with the managing director. There was no longer anyone at Bristol with the will to mastermind the rebuild.

Another home for my charge had to be found. There were no active RAF airfields particularly close to my home, the nearest being those in Oxfordshire. Consequently, I set about examining the likelihood of being granted sanctuary by one of them. Abingdon proved a non-starter, much as I had expected, and disappointingly, Brize Norton considered itself unable to help. I felt it unlikely that Benson, the home of The Queen's Flight, would welcome the presence of my project but, much to my surprise, I received a phone call from Squadron Leader Pete Sowden who proved to be in command of the Engineering Squadron there. He had heard of my predicament and wondered if he could help! 'I have space in my hangar for you, if you want it. It seems to me that you are getting the bum's rush from Northolt. So I thought I'd offer a hand.' His friendly gesture was the one ray of sunlight in a very bleak and disheartening period.

Delighted at last to be able to tell Northolt what to do with its appalling shed, I requested a 'Queen Mary' transporter for the move. I was flabbergasted by the response. It was claimed that none was available! In its stead I was offered a three-ton truck and trailer. I protested, of course, and I'm afraid that I was unable to contain my anger in the process. I started packing everything in polythene sheeting and cardboard boxes once again. I had to ask British Aerospace to hasten the return of the undercarriage legs to enable me to move the fuselage.

One sunny day, we began the insertion of the airframe into the inadequate transport. Our first attempts at lifting the fuselage would have appeared comical to the bystander. We grossly underestimated the weight distribution and, on the first lift, the tail shot high in the sky whilst the front end remained firmly attached to terra firma! Several adjustments of the sling were needed before a successful, near-horizontal lift was achieved. The truck was too short to accommodate the fuselage length. We had to resort to easing the tail into a position immediately behind the cab. Even so, the front of the aircraft protruded beyond the tailgate. Therefore, the structure had to be

On the move. The first attempted lift was not quite right!

laid on beams lagged with horse-hair and felt, the undercarriage legs dangling behind the truck. The wings were placed in the trailer, and all available space crammed with boxes of components.

The journey to Benson was accomplished with only slight damage to the airframe, a remarkable achievement by a caring transport crew, and a success story which far exceeded my most optimistic forecast. Northolt had the final parting shot. Our racking could not be fitted in the truck, and I decided that it should be moved later. Arrangements made, I received a letter from the Group Captain. As his engineers were 'desperately short', he regretted that he could not agree the transfer of the racking — our racking, assembled from discarded scrap — it will be recalled!

On reflection, I thoroughly enjoyed the camaraderie which developed amongst my colleagues, despite seemingly endless, tedious graft performed in abysmal surroundings. I doubt I could have found another body of men so willing to tackle such generally unrewarding toil. Scattered through circumstance, I was alone and faced an uncertain future at Benson. Nonetheless, I was very glad to be rid of Northolt, and Building 120!

9

A non-standard *Gustav*

During the course of the first years of the project, and especially following the acquisition of the *Ersatzteilliste* and the translation of our handbooks by Peter Nolte, Ian and I had discovered that 10639 was not a standard Bf 109 G-2. Equipment variance initially caused much confusion and resulted in speculation about its place on the production line. Many questions remain even as I write. I doubt they will ever be properly explained. Perhaps at this stage of the story, with the aircraft unpacked at Benson I should take the opportunity to describe our findings.

My suspicion that all was not what it might seem was first aroused during my initial inspection at Lyneham, but the significance of my find was not apparent until a good deal later. Inside the rear fuselage, and visible when the radio hatch is removed, is a large alloy label produced by Telefunken denoting the installation of an FuG 7a radio. Its top right corner declares the aircraft as 'Me109F', but a cross had been stamped over the 'F' and the letter 'G' placed next to it. At first I assumed that this was merely old stock being used up, but our stripdown revealed much more. Close by, we found two hydraulic pipes running the length of the fuselage and terminating in blanked unions. These were evidently installed to feed an actuator whose function was to retract the tail-leg, yet, according to the manuals, this was non-retractable in the Bf 109 G. (I have since seen evidence that some early *Gustavs* did have the actuator fitted, but these were exceptional.) Of course, all the *F-series* aircraft were thus equipped as standard.

Just as curious, adjacent to the radio hatch, and bolted athwartships, lies an alloy tube. This forms the front support of the battery tray, but it has a secondary role. Slung underneath lay small compressed-air bottles which were charged via a hatch on the starboard side: their purpose to arm the guns. The *Gustav* bore two such bottles, one for each of the MG 17 machine guns; the MG 151 did not require compressed air. However, our bar carried three bottles as is evidenced by the painted annotations 'MG1, MG2' and in between '*MG FF*'.

A glaring anomaly can be found towards the front. Behind the luggage compartment, the Bf 109 G had a distinctive oxygen system, best described as four rows of coupled spheres, three of which were vertically mounted on the rearward-facing bulkhead whilst the fourth was strapped, horizontally, nearby on the starboard wall. 10639, conversely, is equipped to carry earlier equipment — three conventionally-shaped cylinders.

Proceeding now to the cockpit, beneath the floor lies the complex linkage which transmits control column movement to the respective control surfaces. When we first removed the mechanism, little was known about it but, upon delivery of the first *Handbuch*, we were dismayed to see that it was completely different from that shown in the illustrations. I think we may have suspected a later modification, but there was no physical evidence to support this. The truth dawned when the spare parts list for the *Friedrich* depicted identical equipment.

Taking all the facts into account, I am led to the inescapable conclusion that 10639 had begun construction as a Bf 109 F-3, (the MG FF compressed air points to the sub-series) but had been partly modified to Bf 109 G-2 standards, probably before completion. The most complex part of the adaptation must have been the port cockpit wall. It will be remembered that the 'F' wall incorporated the fuel tank filler neck and its circular hatch, while the 'G' panel moved to the rear fuselage spine. In 10639, this area has been completely reskinned. Bar the insertion of the new fuel filler panel and its associated pipery, the rear fuselage remains pure Bf 109 F, but with some redundant equipment removed. The only external clue to its early origin can be seen (or rather, not seen!) at the extreme rear of the fuselage, on the right side. There, the *Gustav* introduced an oval access panel. It is absent in 10639.

The mainplanes, on the other hand, were early-'G' in construction — with interesting additions! In each, we found a pair of hydraulic hoses feeding from the wingroot and along the front face of the spar into the wheel-well. Strangely, they are then connected to each other, forming an enclosed loop. In other words, they perform no function, yet are still connected to the hydraulic system distribution block mounted on the front bulkhead. During the early development of the mark, trials were conducted with small, outboard undercarriage doors which, on retraction of the legs, mated with the long doors strapped to the legs, thus totally enclosing the wheels. These did nothing to enhance the performance of the aircraft, and production *Gustavs* did not carry them. (They were reintroduced, but only much later in the War, on the Bf 109 K.) Our redundant hoses, therefore, were intended to supply hydraulic pressure to a door actuator via a sequence valve. This last was essential, of course, as the door had to close only after the leg was fully retracted, and open before the leg uplock opened. The valves are still in situ, their unions protected by makeshift cloth covers, and the attachment points for the actuators are evident. (The intended fitment of these doors, incidentally, explains the straight outboard edge of the wheel compartments as distinct from the circular 'F' wells.)

That, then, is the explanation of the installation, but why it was fitted to 10639 is not so easily understood. One possible conclusion is that it may have been used in trials to back the pre-production batch. I feel it is more likely that, together with its close contemporaries, the aircraft was being converted to 'G' configuration, and the new mainplanes built to intended production standard — including wheel doors — whilst trials with the pre-production aircraft were still in progress (or soon after) in order to accelerate the introduction of the new mark into service. Like the fuselage before, the wings were then modified, in their case, by deleting the doors and their actuators.

It is worth recording that the Bf 109 G-0 aircraft bore the *Werk Nummern* 14001 to 14012. Fresh-build 'G' aircraft would all have carried later numbers. Those early machines which bore figures in the 10000 range (and they included Bf 109 G-1s as well as G-2s) were almost certainly assembled using redundant 'F' fuselages. It is my contention that all these aircraft would have been equipped in similar fashion to 10639. Digressing slightly, it is interesting to note that one of the pre-production machines, the Bf 109 G-03, *Werk Nummer* 14003 was formerly a Bf 109 F-4 and was subsequently modified to carry a V-tail.

During its career, 10639 had some of its structure replaced by parts cannibalized from other machines, generally to hasten its repair and return to service. A good example can be found in the horizontal tail surfaces. As we know from my early examination of the aircraft, it was built in Leipzig. The tailplane and elevators, however, originated in Vienna. Alloy labels on each declare 'Wiener Neustädter Flugzeugwerke'. In fact, these were fitted to the aeroplane in November 1942 to replace the damaged originals (see Appendix A). The fairings to the fin bear the *Werk Nummer* 9678, and although such evidence is not borne by the tail surfaces, it is fairly certain that they were removed from the same aircraft, a Bf 109 F.

I have described the differences in internal primer finish found in each mainplane; it follows that one, at least, is not original. The explanation came years later. The aeroplane spent the last year of the War with 1426 Enemy Aircraft Flight operating from Collyweston, Lincolnshire, on trials and demonstration. I was to meet several of the Flight's former members, including one John Westwood, ex-Corporal, and Fitter 2A by trade. 'Lofty' (and he *is* tall!) remembered the arrival of the aircraft and was able to relate that, at the time, they held several 109 wings, all of which were laid out and examined, before the two best were selected for attachment to 10639. The right wing I knew to be original due to some damage on its upper surfaces (again, see Appendix A for the cause). The left wing, therefore, came from another captured machine which must have been built about the same time due to the hydraulic fit already described.

Lesser portions of the aircraft bear incompatible work numbers, the most prominent being the two small cowlings which conceal the engine oil tank immediately behind the propeller. Stamped into the skin of both can be found the figure 7232. They had come from a Bf 109 F-4/B which force-landed at Beachy Head in May 1942. It was repaired and flown, ending its days at Sealand in North Wales. As 10639 was also stored there, it is probable that it inherited the cowls of its slightly older stablemate when its own were mislaid.

Further, at the base of the fin, a small magnesium alloy fairing which covers the tailplane screwjack was also discovered to be foreign. The *W.Nr.* 14249 identifies it as having been removed from an early-production *Gustav*.

Lastly, the large intake for the supercharger is also foreign to the aircraft. It is a genuine Bf 109 G part, but from a later model, probably a Bf 109 G-6. The reason for this assertion is that the upper flange of the assembly should be virtually straight for the G-2. The one now carried is scalloped, the only alteration needed to make it fit the later aircraft because of the introduction of a large bulge immediately above the intake to accommodate the MG 131 machine guns.

Various other items, including the engine, were also changed, as will be apparent when Appendix A is read. The aircraft as it is today, therefore, is very much an amalgam of equipment taken from more severely damaged aircraft. It is representative, nonetheless, of a fighter of the line which survived when less fortunate machines were damaged beyond repair.

A line drawing depicting 10639 and showing some of the main identification features of the Bf 109 G-2 may be found on Page 178.

10

Reconstruction at Benson

Back to the story! Peter Sowden, who was responsible for the overhaul of Hawker-Siddeley Andover and Westland Wessex aircraft, invited me to Benson as projections had indicated a reduction in Wessex work. Days before the move, however, he called me with some bad news. A derelict Wessex was to be delivered which he had been tasked to rebuild on a long-term basis. My floorspace had vanished! He was determined not to let me down and suggested that the move went ahead, while he arranged temporary accommodation for the aircraft. On arrival at its third RAF station in ten years, the 109 was carefully placed in a large double garage which had been hurriedly vacated by the MT (motor transport) section.

My relief at an uneventful transit was soon tempered by worry about the future. From a workforce point-of-view, I was effectively back at 'square one'. The only way to generate interest — and find volunteers — was to lay out the aeroplane for all to see, and wait patiently. Entombed in a garage, there was little hope of attracting a new team. Week followed week during which Benson seemed to accumulate more Wessex — not less! I inspected the aircraft regularly, but it was frustrating to be able to do nothing. On one visit I noticed a slight difference in the garage; someone had built shaped plywood formers and bolted them to my trestles, thus affording the wings cradled support. I soon learned that the officer in charge of the adjacent workshops had thought it a good idea and had instructed his men to make them. I found myself beginning to like Benson!

Pete Sowden was forced to conclude that Wessex work was set to continue unabated and, by dint of some subtle reorganization of his hangar, he succeeded in clearing adequate floor space for the little fighter. One beautiful day in November I began the short move into 'A' Hangar, helped by a number of engineering personnel, all of whom seemed happy to be involved. They would not have realized it, of course, but it was the warmest welcome I had ever received. After three full months of inactivity, once again 'the show was on the road', or at least I hoped it was. Some hours of frantic activity had been needed to effect the move. A few more were consumed by unpacking, but they were all worthwhile. For the first time in years I was set up in a real hangar with light, heating, power, compressed air, telephones and toilets — heaven! It was an *unbelievable* improvement in the situation, akin to rediscovering civilization! Life began to look rosy again.

My next task was to recruit a new team, and in that respect I struck lucky instantly. Taking some pity on my struggles to make sense of the jumble of crates and parts, two friendly NCOs joined me, and before I knew it, the fuselage had been placed on its supports, the wings trestled and the multitude of parts laid out on racking and a couple of cupboards scrounged from the four corners of the hangar. One of the men, Chief Technician John Dixon, ran the hydraulics bay outside which the 109 lay, and he noticed that the main undercarriage legs had 'bottomed', prompting me to relate the reason. I asked whether he could help, and within half an hour, they were removed and on his bench. BAe had rushed the legs to me from Hatfield and time had not allowed a complete overhaul. However, new seals had been fitted, so the first move was to insert hydraulic fluid and pressurize each leg. We did not get that far. Soon after we put fluid in the top, it reappeared at the bottom! There was nothing for it but to dismantle them. This done, it became apparent that the new seals were inadequate, but we were amused to find that the seal pack had been assembled incorrectly, some of the components even being upside down! As I had been unable to find original seals on the Continent, John decided to strip the 'Pilatus legs' and use theirs. (They had been serviced prior to delivery and were, therefore, in first-class condition.) We found some differences internally, but two seal packs were assembled and the 109 legs were soon under test on the rig. I tried to be as helpful as I could in a strange environment, but at one stage I succeeded only in bringing work to a halt.

One leg caused a few problems and John decided to strip it once again. It had to be moved to a bench next door. Now, it is a very heavy item, and, having removed it from the rig, I propped it on the floor in order to adjust my grip. Stupidly, I put my hand inside the scissors torque-link which connects the outside 'sleeve' to the oleo piston, forgetting that the leg was unpressurized. It compressed with lightning rapidity, trapping my fingers. It was very painful, and moving only heightened the agony. Responding to my yells, John came to the rescue. I could tell by the incredulous look exactly what he thought of me — pillock! One finger remained completely numb for over a week (it still bears a mark locating the maximum pressure point) but it was a lesson well-learned. If I was to be involved in the engineering, I had to be very careful. After all, I might damage the aeroplane!

The two legs were installed within weeks of resumption of work, but it was to be some months before we refitted the tail-leg. Although serviced immediately after its companions, several new bolts required to attach it were in the making at Filton. Also, the mainwheels received from the Swiss were excellent but the tyres threadbare, and I had yet to trace suitable replacements.

John's deputy in the hydraulics bay, Sergeant Dave Watson, meantime, decided to investigate the new oxygen system supplied by Drägerwerk.

Supplied in kit form, it needed a deal of design work to accommodate the new components in a very restricted area within the cockpit. Just as important, it had to be done in a way which would result in it looking correct. The regulator, contents gauge and flowmeter are obvious features of the lower, right wall space, all mounted on a hefty alloy panel of distinct shape. I must explain that there was good reason to install a serviceable oxygen system. There were a number of incidents in which carbon monoxide had reportedly penetrated the cockpit, and 10639 itself figured in one. Rather than fit the original, unserviceable system, therefore, I decided that the aircraft must be properly equipped. I recognize that this was a lapse in my avowed aim of accurate restoration, but I intend to fit the correct components when the flying career of the 109 is at an end.

John's enthusiasm led him to take all the remaining hydraulic components off the shelf, one by one, and service them, and he even involved his men when Wessex work permitted. There were frequent interruptions because of a sudden panic requirement for some part or other to be produced at short notice. On these irritating days I returned to my 'trade' — cleaner first-class!

Exposed as I was in the middle of the busy hangar, it became routine to chat to interested passers-by. These exchanges normally lasted but a couple of minutes. One, though, developed into a much more regular event. A young corporal introduced himself as one of the staff from the structures bay, which lay conveniently close to the aeroplane. He asked if he could help in any way, and I suggested that he take a look at the many bits needing attention on the racks. I was impressed when he selected the radiator fairings. They were both in terrible condition, having suffered from deep corrosion, and were extremely complex. Ian Mason and I had dreaded the prospect of tackling them at Northolt where we had no access to suitable facilities. Some weeks later, the first was returned to me — or rather a superb reproduction incorporating some original parts! It was a beautiful piece of work, the first of many from the hands of Corporal Paul Blackah.

Left *A badly corroded radiator fairing.*   **Right** *A Blackah-reconstructed radiator fairing with new radiator inserted.*

I could not believe my good luck. Within a very short space of time I had found two highly enthusiastic craftsmen, several occasional assistants, and all in a very friendly hangar. Coupled to this, I was delighted, one Sunday, to see John Elcome walk through the door! It had not occurred to me that he might like to carry on because of the distance between Benson and his home in Harrow, but there he was, large as life, and ready to help me finish the right wing! He was most welcome. I had been tackling the awkward crevices on my own and had become rather fed up with it. Another welcome returnee was Kevin Thomas; and all three of us made light work of a task begun so long before. It was disappointing that Kevin was unable to carry on because of shiftwork at Odiham, and also transport problems. His car was more in need of attention than the 109!

The first few months in 'A' Hangar were eventful and satisfying. As a result of my continuing typing of begging letters, a variety of rubber and bonded-rubber material came from Dunlop who had also made me new radio mounts the previous year. With this help, we were in a position to be able to install the radiators which were soon to arrive from Stuttgart. One letter introduced me to Malcolm Towse who undertook to produce a veritable list of complicated small parts, including pins, cable-hooks and bolts for the ailerons.

At home one evening I had a telephone call from Lancashire. At the other end was Andy Stewart of British Aerospace, Warton, who had read an appeal of mine in a magazine for some parts for the 109. Although he could not offer anything from the published list, he wondered if he could help. To my shame, I recall saying that I thought everything was in hand, but thank you anyway. Suffice it to say I returned his call within days seeking his help, and this he provided, and continues to provide, unswervingly.

I had succeeded in placing my cockpit instruments with Marconi (soon to be GEC) of Rochester, and a preliminary examination offered every prospect of their return to serviceability. The trouble was, I had no panelling in which to install them! I did have technical drawings, though, supplied by Elmar Wilczek of Aachen. I wondered whether Andy could help. This was far from being a small request. Some time before, I had approached a company in Sheffield, a specialist in this type of product, with a view to manufacturing the two instrument panels and the electrics board and cover. Willing at first, examination of Elmar's drawings revealed that the exercise would have cost £1,400 — in 1983 terms! The company withdrew apologetically.

I expected a similar response from Andy, I must confess, but I had not reckoned on his tenacity and skill. Confidently he set about producing the instrument panels for me. With Andy's help I had overcome a major problem, but a problem which had been wholly avoidable.

When the 109 arrived at Wattisham 22 years previously, it came complete with all its panelling. At Lyneham, I had been dismayed to discover that it had all disappeared and began an investigation. Finally, I tracked down one of the men who had worked on the aircraft. When asked what had happened, he told me that there had been no intention to fit German instruments, so '... we binned them. They were thrown on the rubbish dump'! I was taken aback by the answer. Ian Mason was with me at the time of the phone conversation

The beautiful new electrics panel, circuit breakers exposed.

and he can still recall the mixture of anger and incredulity on my face. Later, I asked Wattisham to search for me, but to no avail unfortunately. I cannot understand the mentality of the people involved in this unfortunate episode. Given custody of a rare aircraft, apparently they had no qualms about destroying its equipment when it could have been put in safe storage; surely a simple task.

Andy and I knew that the manufacture of the instrument panels would be a lengthy job, but we could not have guessed how long it was to take. Shortly after work had started on them, we learned that BAe had agreed to rebuild a Spitfire PR 19 to flying condition for the BBMF. On its arrival at Warton, work on my small parts (and many others) ceased. Three full years elapsed before it was received. The Spitfire was not a popular topic of conversation during this time!

Meanwhile, I had finally found suitable bearings for the wing attachment pins. Of a size no longer made, these are pressed into lugs on the spar centre section and the end of the spar proper. Each wing is then attached by encouraging two large pins through these shell bearings. The left mainplane was complete, bar the pitot system and aileron, and I decided that the time had come to put it in place. John orchestrated the attendance of several 'bodies' one afternoon, and we carefully offered the heavy wing to the fuselage. At first I underestimated the wing dihedral (the positive angle made by the wing to the horizontal when viewed from the front). After several minutes of unsuccessful manipulation, not to mention grunting, there was a distinct danger of an accident, everyone's strength having depleted. The cumbersome mass was hastily replaced on its trestles. It took several more

The instrument panel. Made by BAe and completed by the Team.

attempts before the main pins were inserted. It was a lovely moment for me as I had never seen the aircraft with a wing attached.

The right wing at that time was not sufficiently advanced to be fitted. John and Paul decided, for one thing, that one panel on the lower surface had to be replaced because of an area of scoring. It was not the easiest thing to reproduce, but they used their spare time to good effect. Riveting the new skin brought back memories of the cleaning days, as the same small corners had to be reached through the same small holes. That panel, in fact, was the only piece of skin replaced on either wing, bar a couple of small repairs; testimony to their overall first-class condition. A coat of primer later, and we were juggling the second wing into position.

The Messerschmitt was rapidly taking shape, and I felt content and very confident that we were well on the road towards a successful conclusion to the project. I was even more happy to hear that Ian Mason was returning to the fold! Traditionally, a serviceman is allowed to select his last posting prior to retiring, and Ian had asked for Benson. Not only was this granted, but his new job placed him in our hangar and only a few feet away from the aircraft! On the day of his arrival he found me grovelling under the belly, attempting to fit some new pressure-instrument pipelines. His greeting was in the accustomed form of a caustic remark. It was good to have him back.

He was soon put to work! After attaching the radiator flaps on the starboard wing, we discovered a problem. All our attempts to adjust them to line up with the plain flap failed, and we concluded that the top half, which had been rebuilt, was incorrect and would have to be done again. Not only that, but the plain flap itself, obviously rebuilt before the project, was badly distorted. Not for the first time, we found a depressing case of one step forward, two back. Not daunted, the team de-riveted both. The radiator flap caused much head-scratching as the alterations needed to the alloy sheeting were minute. The plain flap was much easier. Taken apart, they found the old rebuild to have been amateurish. Paul, in particular, was irritated that anyone of his trade should have made such a horrible mess.

The *Gustav* looked like a real aeroplane once more. I was proud of the progress that had been made. I also eagerly anticipated the return of the propeller — from Germany! I had last seen it ten years before, when it departed for British Aerospace Dynamics in Lostock, Lancashire! In common with every other part of the aeroplane, I had to search for the appropriate manual, in this case *L.Dv.514,* an essential document to allow the overhaul of the airscrew. Unfortunately, I had very little luck tracing a copy, and for a long time had to report this fact to BAe. I did manage to arrange the translation of a Finnish document on the VDM propeller, itself taken from German, but neither this nor Farnborough wartime reports provided sufficiently detailed information. Over the years I received conflicting reports on the condition of the propeller, admittedly from different people. An initial inspection had revealed that there was slight corrosion damage evident in each blade and that some bearings were missing from the hub. After dismantling, both these observations were confirmed, plus the fact that each blade tip had been slightly bent. Nonetheless, supplied with the appropriate technical information, it was felt that the unit could be made serviceable.

Later letters revealed that doubts had increased about the possibility of delivering a useful prop to me, even with severe operating limitations imposed, and I was urged to find another! As the only others in the country were fitted to the Messerschmitt Bf 110 G-4 at Hendon, and these I knew to be in poor condition, I could see little hope of salvation in this direction. Following one letter, I began to wonder what had happened to the assembly since its delivery. It stated that the factory had worked near-miracles in straightening the blades and removing the 'extensive corrosion', but there was no hope of returning them to flight condition. Accompanying the letter was a large photograph showing three very bent and corroded blades. The damage shown was consistent with the attached aeroplane having crash-landed, or even ditched! I protested that these were not the blades I had supplied, of course, but I feared for the safety of the real ones. Shortly afterwards, a lengthy missive repeated that in-depth technical information was needed to enable work to be carried out to a standard to suit the Civil Aviation Authority. Also, BAe had experience with British and American assemblies, but had never overhauled a German one. Therefore, they admitted, it was unlikely that they could be of further help.

I turned then to Germany and Hoffmann Propellerwerke which specialized in the manufacture of composite wooden blades. Some of these are used by BBMF but, more interesting, they were fitted to a hybrid Hispano HA1112/Me 109 G operated by Messerschmitt-Bölkow-Blohm.

(Basically a Spanish-built Bf 109 powered by a Rolls-Royce Merlin, MBB had commissioned a rebuild incorporating a DB 605 D engine. I have to say that I thought it not too clever a conversion, most of the new parts being incorrect in appearance. It first flew in April 1982, but its career terminated in dramatic fashion in June the following year. During a take-off from a grass strip the aircraft swung off line, colliding with a tractor and a blast-fence before coming to rest. It was a write-off, but another Hispano was purchased in France, and this time the result, to my eye, is accurate and there is little to betray its Spanish origin.)

I hoped that the Germans had the required expertise to help with the VDM unit, but in a courteous reply the managing director suggested I consider commissioning three composite blades. The cost (then) was £1,000 per blade. There was little chance that I could raise such a sum.

Fast running out of ideas, I thought again of Dowty-Rotol, in all honesty expecting no more than a reaffirmation that the company could not help. Instead, the reply, though guarded, was not negative, and asked permission for one of their engineers to examine the components at Lostock. Days later, examination completed, arrangements were made to transport them to Cheltenham, the prognosis being that a return to health was possible.

I was intrigued some months later to learn from Andy Stewart that one of his MBB colleagues, on a visit to Hoffmann in Rosenheim, had seen my propeller being worked upon! I rang Cheltenham and learned, much to my surprise, that it was indeed in Germany. Dowty had links with Hoffmann and had passed it straight to them!

Some months later it was ready, and Rolls-Royce agreed to retrieve it on one of their regular German truck services. We off-loaded the large triangular box and eased off the lid. Revealed to us was a propeller which I can only describe as immaculate. Hoffmann had even painted it in its correct Luftwaffe colour, *Schwarzgrun* (black-green). It was a magnificent effort all round. On reflection I do not think that I expressed my gratitude adequately, but it was a supreme example of what could be achieved given goodwill and a sympathetic company.

At the other end of the spectrum, I can quote the case of a company in Mansfield which agreed to produce two small, colour-coded knobs for the cockpit. After some years (!), I was told that, 'due to a company restructure', it could not help after all, and my — by now faded — drawings were returned.

An exploded thermostat! A complex little device.

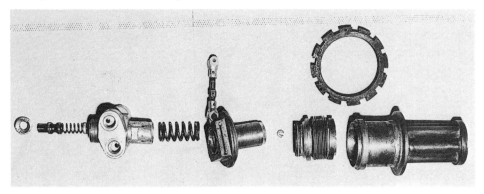

A last look inside the rear fuselage before the fuel tank is installed. In the foreground is the plinth for the compass detector. Above the oxygen bottles protrude, while further back may be seen the empty radio rack.

The propeller was destined to remain in its distinctive box pending the availability of the engine. But, in the meantime, we had lots more work to do. John had made startling advances with the hydraulic system to the extent that almost everything was in place. On the lower front bulkhead of the fuselage lies a large housing inserted into the engine coolant system. Within this is a

Under the belly. Two coolant cut-off valves and fuel filters stand out in a very congested area.

thermostat of mainly brass construction. We had experimented with it and discovered that it was not functioning. It had to be dismantled. Easily said, but far from easily done, as we discovered.

We removed every screw and locating key that we could find, yet still we could not take the body apart. The only remaining barrier appeared to be some solder, so we marched across to station workshops and persuaded the head man to melt the substance by applying a flame to it. A minute or so later there was a tremendous bang as the thermostat separated and its parts various hurtled around the shop! Fortunately no-one was hurt, but it took some time to retrieve all the bits. The reason for the explosion became obvious when we examined the innards. In the bottom of the receptacle there nestled a dribble of clear fluid. Evidently inflammable, it had not taken too kindly to having intense heat applied to it! The cause of the unserviceability was deep corrosion on a stainless-steel stem. Another job for Malcolm Towse!

Once more a begging letter found its way to a British company, in this instance, Negretti Aviation of Hemel Hempstead. Arthur Gradding, on the point of retiring, volunteered to rework the maltreated thermostat. A suitable capsule (essential to its operation) was found and, after assembly, it was subjected to a test programme before being returned. Fitted with its new stem, it was inserted into its housing, and the hydraulic system was complete.

Still on hold because of the Falklands War was work on our fuel system at Wolverhampton. Two years after Ian and I had delivered it I decided to

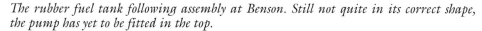

The rubber fuel tank following assembly at Benson. Still not quite in its correct shape, the pump has yet to be fitted in the top.

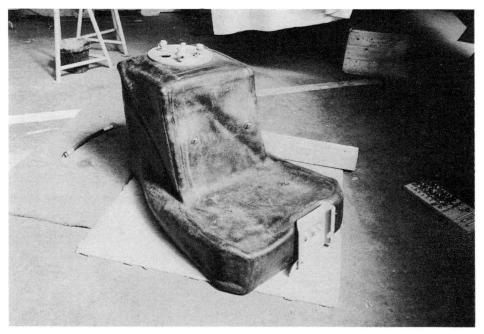

retrieve it, as the fuselage was in a condition to accept it following the fitment of the armour plating and plywood cladding around the tank bay walls.

Regrettably, very little had been achieved during its absence. We were appalled to find that the tank had been allowed to collapse under its own weight, despite my explicit instructions that it had to be supported to prevent possible damage. Further, the rubber base bore evidence that it had been standing on a wet surface. Unimpressed, we decided to have a look inside even though the pipework had been installed. We found that it had been incorrectly assembled and none of the pipe-unions had been wire-locked. Worse, the bottom of the tank was covered in metal swarf, and we even found a broken drill-bit! It had taken two years to accomplish nothing.

Within two weeks the tank was ready for installation, but there was one remaining problem. We had been unable to induce it to return to its correct shape, as a result of which, the long fuel pump, dipstick and contents transmitter could not be fitted. Therefore, we had to resort to installing the tank incomplete, hoping that its suspended weight would restore its profile. To an extent, it worked, but a good helping of persuasion was needed to complete the task. Fitting hoses to both tank groups was a real trial of skill and patience. Paul came as close to emitting the odd expletive as he has ever been; John, Ian and I were not nearly as restrained! In went the fuel pump, which had been beautifully overhauled by Plessey, and the other components. The airframe fuel system was now complete.

Fuel tank in place, the team turned its attention to the belly panelling. It will be remembered that the exchange deal I arranged with the Finns gained several panels lost from the aircraft since the War. These had been copied

A largely rebuilt airframe awaiting the right wing-tip and tail control surfaces.

from the Bf 109 G-6 aircraft on display at Tikkakoski. One panel required a complete rebuild as it could not be made to fit 10639, but missing from it were two Dzus-type fasteners which our Finnish friends had been unable to supply. In a remarkable quirk of fate, an old acquaintance of mine, Dick Melton, had asked to examine the flap drive of the *Gustav* some weeks before. He was rebuilding a Hispano at the time and, for some reason, the entire system was missing. After some hours of looking, measuring and asking questions, he asked if he could return the favour in any way. Paul mentioned the fasteners which were holding him up, whereupon Dick offered us some! Apparently, this was yet another 109 part used in the Pilatus P-2, and he had a stock of spares to support one such machine.

The last Finnish panelling to arrive had been the two long strips which fair the top surface of the wing to the fuselage. The supply of these was stipulated in exchange for the Merlin engine. Due to movements of personnel within the Finnish Air Force, I had discovered that I was suddenly corresponding with an officer who knew nothing of the exchange details! Many letters, and months, later, he was finally able to establish the validity of my claim, and set about having the fairings made. Too complicated to be produced by the Museum, he enlisted the aid of Valmet, the State aircraft factory, and the resultant panels had the appearance of the originals.

11

The engine returns

Ibegan to receive strong indications from John Rumbelow during the autumn of 1987 that the engine would soon be ready, and this prompted the Benson team to turn its attention to the cowlings. The top two had already been repaired by the 'old team' at Northolt, but further work was needed on the catches which showed signs of brutal treatment over the years. A new hinge pin was acquired and the two halves fitted with a new rubber seal of the type resistant to the effects of heat and oil.

The lower cowl had also received attention some years before, on this occasion by the personnel of 71 Maintenance Unit at RAF Bicester, when some tears in the skin of the fairing enclosing the oil cooler had been welded. To complete it, we needed only to fit a bonded rubber strip, part of a length made for us by Dunlop; but that was far from the end of the story. The large cowl, carrying the heavy cooler, is hinged at its right edge; the other half of the hinge being integral with a long channel constructed of poor quality steel. This, in turn, is mounted to the engine by placing it over 24 studs on to which the exhaust ejector stubs are secured. Both this channel and its brother were in very poor condition, corrosion having consumed some vital areas — Sod's Law, I suppose! John Elcome and I cleaned all the accumulated muck from them (and there was a lot of it) and then, using abrasive paper, eliminated the lesser areas of rust. With the worst portions properly exposed, John Dixon devised repair schemes, cut out the offending material and formed replacement sections. The next step was to follow our well-worn path to the door of the station workshops to ask for some careful welding. Once again, as I had no authority to request assistance, our channelling lay on the shelf for many weeks, awaiting a lull in official work. When I next approached the flight sergeant in charge, I was able to claim truthfully that they were most urgently needed. The engine had arrived!

In early February 1988, a Rolls-Royce lorry arrived bearing a huge yellow box. The DB 605 was a large engine, I remember thinking, but not that big! In fact, no smaller crate could be found at Filton and they were forced to

The DB 605 A nearly finished at Filton. The twin camshafts are exposed.

adapt one designed to enclose a vectored-thrust Pegasus jet engine. It was fork-lifted into the hangar, whereupon we eagerly undid the catches and lifted the lid. Inside, in an envelope of clear polythene, the engine gleamed. It looked better than new — a really breath-taking sight.

I had agreed with John and Roger Slade that we would not begin attaching the engine until they were able to visit Benson. Two weeks later, the great day arrived. That Sunday morning I arrived at the appointed hour with my son, Graeme, and was amazed to find the team already there, together with John, Roger and several curious spectators, Service and civilian. How some of them had known our intended task that day was beyond me, but I

The massive fuel injector pump and some of its pipe-work in position.

The top plate of a lovely thoroughbred engine.

could have done without the audience. The engine had already been unwrapped, and we prepared it, and the airframe, for mating. Because the space at our disposal was confined, the only way we could manoeuvre the motor was to employ a nearby winch built into the hangar roof. Therefore, we moved the airframe close to it and raised the tail to bring it into a suitable attitude. Or, rather, we tried to bring the tail to a suitable attitude!

The aircraft, even minus the engine, is very tail heavy when sitting on its undercarriage. Moreover, the low-thrust line of the Daimler-Benz inverted engine implies that, in horizontal flight, the empennage sits high relative to the nose. To reproduce the horizontal in the hangar, we had to lift the tail

The Daimler-Benz ready for delivery, but as yet devoid of its external pipework.

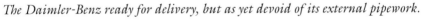

The beautiful DB 605, hoist beam attached, awaiting mating to the airframe.

higher and higher until the situation became precarious. At the other end, meantime, John and Paul bolted to the rear of the engine a heavy frame, the mechanism for attaching the MG 151 cannon. Lo and behold, once attached the centre of gravity of the mass had altered significantly and the whole adopted a 'tail-low' position. It was then that the rest of us realized we had been wasting our time raising the tail so extravagantly! A few moments of embarrassed laughter and shouts of abuse served to make light of all our unnecessary effort, and we were finally ready to attach our lovely engine.

Not surprisingly, since the bearers had been dismantled for inspection, we

After hours of work, the engine is in position.

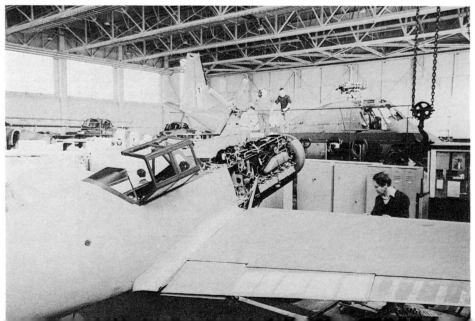

Some hydraulic hoses replaced, work con-
tinues to complete the engine.

Unusual view of the Bf 109 G showing the
asymmetric rudder.

found that the lugs were incorrectly set. Back and forth went the heavy lump to allow a series of adjustments to be made. Every movement was made extremely slowly and carefully, lest we cause some damage. The process took some hours to accomplish, during which time most of our audience vanished, presumably through boredom. I, for one, was not sorry!

Eventually, late in the afternoon, all four pins were inserted and secured, and, with bated breath, we removed the hoist. We had weighted the rear fuselage, but there was a worry that perhaps not enough weight had been applied. With the fuselage in a tail-up attitude (albeit less exaggerated than at first), had the mass of the engine proved too much, I doubt whether we could have avoided a nasty accident. But our calculations had been adequate, and the aeroplane remained rock steady. For the first time in 15 years the engine was 'home'.

Only Ian and I, though, could appreciate the contrast of the sight with that of 1972. At Lyneham, the engine had been scruffy and covered with a filthy concoction of oil and preservative; the airframe damaged and clad in an untidy coat of paint. The picture before us could have been taken from a Messerschmitt production line of 1942. Not unexpectedly, drastically altered in appearance by the sudden addition of the DB 605, the aeroplane was the focus of attention for many days following. I gained the impression from their reaction that the hangar staff had doubted whether the project would be successful, but the engine changed all that.

Roger Slade and Russell Stokes (his assistant at Filton) had been placed under some pressure to remove the engine from its small shed over the previous months. I believe that, as a cost-cutting exercise, Rolls wanted the workshop closed to reduce the heating bill, etc. As a result, a lot of ancillary work remained unfinished. Most noticeably, almost all the external pipework was missing. Comprising rubber hosing of various diameters and alloy tubes, the former presented few problems as we had various sources of suitable material. It was simply a matter of cutting it to the lengths indicated in the *Ersatzteilliste* and coupling the alloy end fittings. However, there was some trial and error, as several quoted lengths proved too short or, as in one extra-ordinary case, far too long! The alloy tubing was a different matter altogeth-er. Some time earlier I had 'persuaded' Avica of Hemel Hempstead to pro-duce several large bore pipes for the wing radiators — the originals either missing or having been mangled by the frequent dismantlings over the years. It was unkind of me to 'hit' them twice, and they felt unable to help further.

I could find no other specialist in the field who was willing to assist, so once more I turned to Andy Stewart. It was vital that the new pipes lay cor-rectly. In common with most piston engines, little space was afforded by the close-fitting cowlings. If I made even a slight error, therefore, it would have involved Andy in a lot more work. There followed weeks of intensive study of engine photographs to assess the fit of each individual pipe. I decided to use thick copper wire to trace each path, and whilst we achieved a remark-able success rate, it was a difficult exercise. Even the slightest tweak at one end of a wire caused an enormous displacement at the other. Sometimes, several pairs of hands were needed to form one small length of wire. Complex pipes at the front end of the engine required several journeys between Benson and the workbench of Roger Slade at Bristol before a satis-factory fit was achieved.

While I 'designed' the new alloy pipery, rubber hoses were being rapidly assembled and fitted by my colleagues. It all seemed a little unfair; it took me hours bending a copper wire, yet the others seemed to be fitting hoses as though they had been lying on the shelf! Thanks to Andy's speedy efforts, the engine soon had most of its complement of external equipment and we started to fit the cowls. First to be attached was the lower, complete with its heavy cooler, followed by the upper and lower halves at the front, enclosing the annular oil tank.

Things were going really well, when the RAF posting system intervened. A new warrant officer took charge in the hangar and, as appears to be nor-mal, a reorganization was soon ordered. The first move was a real shock. He decided that John, being the senior NCO in the hangar, was wasted in the hydraulic bay, and he was moved to the floor to take charge of one of the Wessex overhaul squads, replacing Ian who had a few weeks to serve before leaving the Service. Overnight he became heavily embroiled in helicopters, and the 109 was forced to take a back seat. The second change to affect us was a major relocation of a large proportion of the hangar itself, and the aeroplane was moved to a corner which offered only one advantage; it was closer to John! They say that things happen in threes. Our third event of the period was when Paul did a realistic impression of Barry Sheene. Going round a corner on his beloved motorcycle, he hit a rough patch of road and parted company with his steed whilst still going at a fair lick. He broke one arm in four places.

John found that his new job occupied most of his working day. The time allowed for an overhaul of the ageing Wessex always seemed to me to be ridiculously short, and his main battle was against the clock. On the days that I attended Benson, I could no longer guarantee that he would be available. Indeed, it became normal for me to see him only during his short breaks and at lunchtime. I wondered what I had done to offend the Air Force! Coupled to this, of course, I lost Paul whilst he recuperated from his spill. I was amazed that he actually put in the occasional appearance, the first only days after the accident, although looking very battered and frail.

It was a good time to make some headway in the cockpit. Over the years of the project, one of the scarcest 'commodities' had been the services of a friendly electrician. At Northolt, only Pete Hayward had come forward, but he was able to play no part in the reconstruction before leaving to become a civilian. To most of us, electrics were a real mystery; a tradesman had to be found. Luckily, one such, Eric Faithfull by name, offered to help when he could. He had been cajoled into helping earlier. We had inserted a new-build pitot system in the port wing, incorporating a pitot-head that Peter Nolte obtained for me in Germany. Since he was the only person in the hangar qualified to test it, I was compelled to ask for his time, which he readily gave. The system proved to be a leaky disaster as first fitted and, as work on the engine became a priority, it was abandoned for some time.

Most of the cables installed in the cockpit were intended to supply the engine and instruments. They all routed to their respective terminals in the distribution panel located on the starboard cockpit wall. The original, as I have related, had been 'discarded' at Wattisham and, being a very complicated piece of equipment, I was not optimistic that I would be able to reproduce it. I had many pictures of it, but these were hardly adequate for the purpose. The obvious solution would have been to borrow a panel for copying, but this was impractical. It would have been a major undertaking to remove one from an aeroplane, and I knew that the museums would be most reluctant to provide such assistance. However, perhaps I could persuade them to help with drawings and measurements. My more recent correspondence with the Finnish Air Force Museum clearly indicated that I could expect no further co-operation. They sought fittings for their Hurricane in exchange for anything that I might ask, and I'm afraid I had no hope of finding such rare parts.

One other contact, I felt, might be in a position to help. Richard Lutz was involved in restoring some of Ed Maloney's aircraft in the Planes of Fame Museum at Chino, California. He had helped me with information in the past, derived principally from his work on a Bf 109 G-10, *Werk Nummer* 611943. This aeroplane, incidentally, could be restored to airworthiness quite easily, if only a propeller and spinner could be located. I believe Maloney would like to put it back in the air, but there are so many other aircraft to tend that it has only been a dream to date. I digress, yet again!

The Chino aircraft still carries its electrics board, and I asked Richard to provide me with some measurements and information. I would have loved to have borrowed the thing, but that was out of the question. Instead, I decided to make a panel based on all the details I could find. As usual Richard obliged, but the requested details were never used, as a complete set of drawings was loaned me by Elmar Wilczek of the *Technische Hochschule* of Aachen. Many, many sheets of technical drawings were needed to present a full picture of a panel of under two feet square in size, and it took me hours, with them all spread out on the floor around me, to begin to understand them. I had persuaded many companies and people to assist me in the past, but I really could not see anyone being daft enough to help with this! In truth, I didn't. It was Andy Stewart who suggested to Jack Green of Flight Refuelling Limited that he might care to tackle it. I have often wondered if he knew what he was taking on. Whether he did or not, he stuck with it, bless his heart.

The panel comprises a base plate onto which are mounted a selection of circuit breakers and terminal blocks. A smaller, central control panel holds further CBs, a lighting rheostat and a pitot heat indicator. Over the top lies a cover of distinctive shape through which the buttons of the breakers protrude. Before the drawings and Jack Green arrived, I had found a couple of circuit breakers in the RAF Museum store at Henlow. The remainder came, once again, from Peter Nolte, some still in their wartime cardboard boxes! He also gave me an unused rheostat and pitot-heat indicator. The terminal blocks had come from Cardington. Apparently, someone had turned up at the gate one day and deposited a small cardboard box full of German electrical equipment, asking: 'Can anyone make use of any of this?'! Whoever you are, Sir, many thanks! I wish there had been more like you.

The circuit breakers in the Bf 109 function as switches, as well as fuses. They connect and disconnect power to the selected services. It was vital that they were properly examined and tested. I trudged across to the avionics workshop and sought help, and some weeks later a telephone call revealed most to be serviceable. Not surprisingly, it took months to fabricate our new panel, but when the time came for it to be wired up, it was available to us. However, Eric, like John, was in great demand in the hangar and he was only able to do a few minutes work every now and then. When you are trying to make sense of complicated German wiring diagrams, it is advisable to be able to devote your undivided attention. For the first time I received official help.

Appreciating the situation, RAF Abingdon released two electricians for one working week. In that space of time, most of the panel was completed — but not quite all. I had to wait, then, for Eric's workload to ease, whereupon he tried to complete the wiring left by the tradesmen. While most of it proved acceptable, there were a few errors and, once again, it was one step forward and two steps back while everything was double-checked. The mutterings emanating from the cockpit while Eric scrutinized the diagrams, were thankfully not intelligible, but they were amazingly reminiscent of Ian Mason in full flow — but with the volume turned down!

I was relieved to learn that we had inserted the correct number, length and type of cabling to supply the board. Those towards the front of the cockpit would have presented little problem to replace, but from the rear, wiring to connect the compass, radio, and battery passed through the fuel tank bay en route. Replacing any of these, therefore, would have involved removing the fuel tank — again!

Everything was moving depressingly slowly. In the preceding months, the Messerschmitt had been substantially reconstructed and, barring the cockpit, only detail work remained. But suddenly my new team had been effectively withdrawn. Another concern I had at the time was that the Ministry of Defence became embroiled in deciding the future of the aircraft. Of course, although I started the project with the aim of making it airworthy, there was no guarantee that its 'owners' would allow it to fly. I suspect that no-one in London thought that I would succeed, but recent progress had changed that opinion.

12

Politics

The year 1989 proved to be the most unpleasant of the entire project, but, before describing the sorry tale, I must relate an incident which occurred soon after my arrival at Benson.

I sent a letter to the Chairman of the Historic Aircraft Committee in Whitehall to keep him abreast of my work. Weeks later, I received a reply which rocked me back on my heels. It informed me that my project had been discussed at a full session of the Committee and that they 'were disappointed to learn that the aircraft would be unable to get an airworthiness ticket again'! Apparently a Kent company had offered to extend a display building at Manston, already housing a Spitfire, to accommodate the Bf 109. Therefore, could I give some indication of the anticipated date of completion! I carefully re-read the letter I had sent, but nowhere had I suggested that I would be unable to complete my work. Within minutes I typed a letter to refute the claim and I heard no more of it. (Indeed, a subsequent HAC Chairman denied that the plan had ever been discussed.) But who gave birth to such a malicious rumour?

Immediately following this isolated incident, I was reassured of the support of the Committee, and a meeting at Benson was suggested to discuss the future, particularly with regard to the requirements to meet an airworthiness certificate. We had always worked on the premise that the 109 would be operated by the Royal Air Force, and all our work records were kept on military job cards, but there had always been a possibility that a Permit to Fly would have to be sought from the Civil Aviation Authority. For one reason or another, the proposed meeting did not take place.

Barring occasional reports and acknowledgements from MoD, little occurred — officially — for two years. Unofficially, rumours were rife! From various sources, we heard that our 'baby' was to be sold. Furthermore, during an evening visit to bolster the profits of his local hostelry, Paul Blackah heard a lady, who had direct family links to the historic aircraft scene, state that a bid had been made for the 109. All this was very worrying, as might be appreciated, but I have never been able to learn whether there was any truth in either rumour, or any of the others which abounded at the time.

A new HAC Chairman, Group Captain Appleyard, took post and requested the meeting suggested by his predecessor. Early in 1987 I gave him a guided tour, together with Len Woodgate, the Curator of the RAF Aerospace Museum at Cosford, and Wing Commander Paul Brindley, who was in charge of the Historic Aircraft Collection at St Athan. The purpose of their visit was to assess my work, and I was gratified to receive a great deal of complimentary comment from all three. Paul, however, was convinced that the aircraft would not be allowed to fly — which made me even more determined! Nevertheless, having realized that my work would culminate in a perfectly airworthy machine, all three had to recommend a scheme through which it could be operated. Len proposed that it be flown by the RAF Museum at Cosford, funding the initial operation through gate takings. The idea was to assess feasibility by flying it for one season, and it was submitted to the HAC for discussion.

For the next six months, although Len and I kept in constant touch, nothing happened. Later, I received a letter from Paul Brindley disclosing that the HAC 'was in abeyance' pending a review of the Historic Aircraft policy. (I believe that this was the result of realization that the RAF Museum was unlikely to be able to meet its financial obligations in repaying the vast loan arranged to build the new Bomber Command Hall. It resulted in many changes at Hendon, and the demise of the St Athan collection, not to mention the disposal of a number of old aircraft.)

A further six months elapsed. I was anxious that I should begin to involve the CAA. If the reconstruction advanced too far, I thought it highly likely that an inspector would demand some dismantling for inspection purposes, and I preferred to avoid the possibility. In an attempt to accelerate a decision, I sent a letter supporting the 'Cosford Plan' to Air Chief Marshal Sir Patrick Hine, whose office as the Air Member for Supply and Organization (AMSO) encompassed the HAC, and the following month attended a meeting in his office in Whitehall in the company of Dr Michael Fopp, the Director of the RAF Museum and yet another HAC Chairman.

There were a few surprises that day. The most astounding was that, far from giving vocal support to the Museum flying the 109 as I had been led to expect, it was obvious that Michael had abandoned the idea. The plan, I was told, had always been a non-starter as the Museum was not established to fly its exhibits. My team and I had liaised with Len Woodgate for a year, even visiting Cosford to discuss our requirements. We had all agreed with the plan and it seemed to us all that displaying the Bf 109 would repay the help given by the Museum to the project. It was a great disappointment to learn that it had been discarded, but overriding everything was some anger that no-one had bothered to tell us many months before.

The main purpose of the meeting became the formulation of a new plan for the future operation of the aircraft which would be put before the Air Force Board for approval. Sir Patrick favoured operation by BBMF, whilst I laid out a case for independent operation, based at Benson. Both were to be investigated.

There followed a succession of visits to the aircraft by staff from Strike Command, including the engineering authority with responsibility for BBMF. The Flight itself had little desire to add the Messerschmitt to its fleet, a Daimler-Benz engine, for example, being an unknown quantity. Besides, the finance provided for operation of the Lancaster, Spitfires and Hurricanes was meagre. To fly the 109 in addition would probably have entailed removing a Spitfire from service (possibly even the newly-rebuilt PS915) to make ends meet. A compromise solution allowed Benson to remain its home base and servicing to continue to be in the hands of my team. We would, however, be governed by BBMF in matters engineering. To this we had no objection, but none of us could understand an attached proposal — that winter overhauls would be performed at Coningsby, which would have entailed all of us decamping to Lincolnshire! Whilst supporting the overall plan, I registered our strong objection to this last item. Later, I was given a brief glimpse of the final plot, most of which, I remember, met our requirements. The bit I had qualms about was the cash figure requested to cover annual costs which was, in my opinion, overly inflated. I reasoned that a far lesser figure would have sufficed to 'kick-start' display flying.

The Standing Committee of the Air Force Board met towards the end of February 1989 and discussed the proposal formulated by the HAC Chairman. There followed an inexplicable silence. In fact three weeks were to elapse before I was invited to see the result of the Whitehall deliberations. (I had expected a copy, since the decisions taken would affect myself and the team directly, but I was not to be so privileged.) In the silence of the Station Commander's office at Benson, I read the statement — and then re-read it, because I could not believe the contents! The Air Force Board had decided that it would be 'inappropriate' for the Royal Air Force to be seen operating a former enemy aircraft. (A letter elaborated some time later that the decision had been 'due to the sensitivities which still existed about the last War.') However, in recognition of the effort put into the aircraft, it was felt that the 109 should be flown, and directed that it be offered on loan to the Imperial War Museum! (A similar suggestion had been made during the meeting with Sir Patrick, but rejected because, like the RAF Museum, IWM did not fly any of its exhibits.) Shown to me at the same time was a Press notice which announced the transfer of the Messerschmitt to Duxford. It also stated that the people responsible for its restoration would be following it there! Intended for release some days earlier, it had been withheld on the intervention of the Station Commander as he knew that I had not been shown it. If this was not bad enough, a meeting had already been arranged to discuss handing over the aircraft. Due to the exceedingly short notice given, I was unable to reorganize my flying roster and could not attend. I was *very* angry.

In effect, before I was permitted to know the findings of the Air Force Board, the plan to operate the aircraft within Strike Command had been rejected, it had been offered to IWM and provisionally accepted, and a Press release prepared. Whilst the Board had recognized the achievements of the team, it obviously did not feel that we deserved any say in the future of 'our' aeroplane, nor did it appreciate our desire (right?) to be closely involved in its eventual operation. I was annoyed by the cold-blooded insensitivity of it all.

At the meeting, objections were raised on my behalf by the Station Commander and transmitted to MoD. Nevertheless, a few days later I was informed that the transfer would take place, and the Press statement, slightly amended, was issued. It still declared that the restoration team would be going to Duxford!

I was asked, then, to give an estimate for completion of my work in order that a transfer into the hands of a Duxford-based engineer could be arranged. I refused to comply. I would have nothing to do with any plan which saw the 109 pass from the hands of the men who knew it to someone who had no knowledge whatsoever of the type. In response to a heavy hint that transport would be arranged to move everything regardless, I called a halt to work in the hangar. There was no point in continuing the restoration if the aircraft was to be taken from us. Lest the threat was followed through, I extracted all the equipment which belonged to me, together with all the manuals, put them in my car and drove home.

No work was done, nor did I see Benson, for some two months. This is not to say that nothing happened. I was regularly harangued over the telephone that I had no option but to comply with the arrangements. Counterproposals, such as operation through sponsorship, got no further than the ears of a very deaf Chairman. Apart from my objection in principle to passing the 109 to someone else, Duxford was a long way from the homes of most of my team, most notably the Bristol boys. A road journey would have taken them in excess of three hours each way, and it was manifestly impractical to expect them to undertake such on a frequent and regular basis.

I was later asked to attend a meeting at Benson and agreed, provided the purpose was to explore alternative plans with IWM. On the day I was dismayed to discover the sole reason why the Duxford staff had made the journey had been to begin arranging the transfer of the aircraft. I was far from impressed by the subterfuge and let it be known. For the benefit of the visitors, I explained my objections, both to the suggested move and to the handover to another engineer. John Rumbelow, who had been listening quietly to the exchanges which ensued, interjected with his own solution to the impasse. Why did we not divide the year between Benson and Duxford? Depending on display venues, we could surely devise a fair allocation which would reduce the number of road journeys to far off Duxford. IWM, however, could not agree. As they were expected to bear all the costs, not surprisingly they insisted that the aircraft remained at Duxford. I closed an upsetting meeting by suggesting that they would have to complete the aircraft themselves.

Many weeks passed during which only occasional work was performed on the aircraft. I asked my contacts in industry to defer further work on several parts that were in the making. It was extremely demoralizing and worrying. There was, after all, a very real prospect that I would not be able to complete my project. I received a string of calls from the HAC Chairman, the first of which demanded a final decision from me as to whether I would agree to complete my work. I repeated that my team had to be given responsibility for the maintenance of the 109 — and at both airfields. These were my minimum requirements. Successive calls posed the same question.

It was apparent that he did not believe, or would not accept, my answer.

The stalemate endured for several months more. The HAC Chairman was busy in the background. Several of my contacts reported that a certain Group Captain had called enquiring about the availability of this or that. In Bristol, poor John Rumbelow did not know what was happening! The previous year (before the Board met) he had suggested that it would be prudent to overhaul another motor as spare. One was subsequently found, but the future of the aircraft decided, it was highly unlikely that any of the Rolls-Royce team would be able to tend the installed engine, let alone work on another. After years of work, they, too, were being denied a future they had eagerly anticipated. Yet, MoD still hounded John to rebuild another engine — free-of-charge!

Finally, in October 1989, I received the first indication that the 'Rumbelow Plan' might be acceptable. Several telephone calls served to draft an amended agreement which included my conditions. I called everyone concerned with the good news and we arranged a resumption of work. But all was not as it had seemed. In November, a new Group Captain assumed the chair. Colin Cheesman discovered that nothing at all had been agreed, despite the intimations of his predecessor. After a good deal of preparatory work, he called his first meeting at the beginning of March 1990.

I took with me a list of demands, all agreed with the team, each of which I decided must be agreed if we were to proceed. With the painful experience of the previous months still burning in my memory, I anticipated a brief, bruising encounter. In fact the meeting (attended also by Group Captain Reg Hallam, the chief pilot-designate) lasted three full hours during which Colin raised all the topics I had listed. These were calmly discussed and summarily agreed. It was evident that he had diplomatically liaised with Duxford beforehand. It was the first of many meetings with Colin, all of which were conducted in a friendly and considerate manner. I could only wish that he had been appointed a couple of years earlier.

It had taken one full year of depressing argument and occasional acrimony to unravel an unfortunate situation, all of which could have been avoided through a modicum of prior consultation and common sense. The Imperial War Museum had been as much a victim as had been the restoration team. Offered the loan of a unique aeroplane, it readily accepted, but had no inkling that a hostile reaction from the Benson team would result. Recognizing the determination — and right — of my colleagues to continue caring for the Bf 109, it agreed a compromise. It was a generous gesture, and it augured well for the future, but a future that was by no means ideal in the eyes of either party, especially the team. We were all relieved to be able to return to work knowing that our future, and that of 10639, was secure albeit in a form we had not envisaged or wished.

13

Awakening a dormant engine

One essential item was still missing from the rear fuselage — the battery. There was little point, of course, in seeking an original unit, so a modern substitute had to be found. I had discovered the required weight and also the power output and, armed with these details, approached every manufacturer known to me. Surprisingly, most cells in production were far heavier than the German, or dimensionally unacceptable. (Ironically, a heavier battery would have proved useful, as we later discovered when checking the weight and balance of the fighter!) The only suitable unit was made in America by Gill, and the UK distributor offered one free-of-charge. It was suggested, though, that I deferred collection until it was needed. It was sensible advice, but I rued the day I listened to it as my would-be benefactor had gone out of business by the time I asked for it! The new distributing company could only offer a discount on list price, and I requested Colin Cheesman to arrange a purchase, rather than find another supplier.

Paul Blackah and John Dixon constructed a new framework for the battery tray. Yet another part of the airframe which had vanished since the War. It had taken me some years to find enough information for me to feel confident that one could be reproduced. It is in the form of a pair of rails supported in the fuselage, within which lies a simple felt-lined tray. The battery is inserted through the radio hatch and placed onto the tray which is then slid rearwards into position under a restraining clamp.

The compass system returned after a thorough overhaul by Ferranti in Edinburgh, and we installed the *mutterkompass* ('mother compass' — the detector unit) on its pedestal immediately behind the fuel tank. Nearby we attached an aerial for the new Becker radio, but not in a position intended for a communications radio. The original, a FuG 7a, was a High Frequency (HF) rig and employed a long cable antenna suspended between the fin and a mast mounted on the rear portion of the canopy. This was of little use to a modern VHF set, but, as it happens, the Bf 109 G was also equipped to carry a device referred to as FuG 25. This was a basic IFF (Identification,

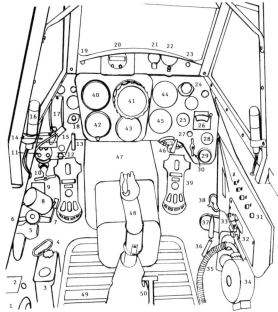

The cockpit of Black 6.

1. Tailplane trim wheel.
2. Tailplane incidence indicator.
3. Priming fuel tank and pump.
4. Port radiator isolation handle.
5. Throttle friction knob.
6. Fuel cock.
7. Fuel injector cut-off lever.
8. Throttle grip.
9. Mechanical undercarriage position indicator.
10. Supercharger filter control.
11. Electrical undercarriage position display.
12. Undercarriage select push-buttons.
13. Spark plug cleaner handle.
14. Engine start switch (under guard).
15. Canopy jettison lever.
16. Instrument panel light.
17. Magneto switch.
18. Battery cut-off button.
19. Firing dolls-eye for port underwing MG 151.
20. Becker radio (location of ammunition counter display).
21. Standby compass (normal location of Revi C/12D gunsight).
22. Clock.
23. Firing dolls-eye for starboard underwing MG 151
24. Electric socket for Revi gunsight.
25. Propeller pitch indicator.
26. Temperature gauge, coolant and oil.
27. Fuel low-level warning light.
28. Fuel contents gauge.
29. Pressure gauge, fuel and oil.
30. Handle for undercarriage emergency lowering.
31. Circuit breakers.
32. Panel lighting rheostat.
33. Pitot-head heater indicator.
34. Oxygen regulator.
35. Oxygen panel, contents and flow.
36. External stores jettison handle.
37. Radiator flaps operating switch.
38. Starboard radiator isolation handle.
39. Rudder pedal.
40. Compass repeater.
41. Artificial horizon.
42. Altimeter.
43. Air speed indicator.
44. Boost pressure gauge.
45. Rpm gauge.
46. Windscreen de-ice valve.
47. Cover of MG 151.
48. Control column.
49. Raised floor plan.
50. Control column lock.

10639 awaits its propeller. Speedy access to the engine is obvious.

Friend or Foe), and employed a small whip aerial fitted through a hole in the lower fuselage. The new aerial put in this position did not look out of place. (Very few machines actually carried the full IFF layout. Indeed there is evidence to suggest that only those aircraft operating in defence of Berlin used the equipment.)

Andy Stewart continued to be helpful in the background. John badly needed to check the hydraulics now that the system had been completed, and especially the operation of the undercarriage. But I had an additional worry. The mainwheels supplied by the Swiss Air Force Museum proved to be in beautiful condition but, as I have said, they had arrived clad in very worn tyres. Through Andy we learned that the closest modern equivalent was a cover as used by the Canberra jet bomber on its nosewheels, and the Avro Shackleton on its tailwheels. Two were offered by the Goodyear company and at Heathrow they kindly fitted them to the wheels while I waited. Their size was not exactly correct and I feared that the space in the wings would prove inadequate. Andy arranged the loan of a hydraulic test rig from Warton which we coupled to the 109. It was a marvellous machine which we

Undercarriage retraction test. Paul Blackah, back to camera, pumps vigorously while John Dixon watches. RHIP—Rank Hath Its Privileges!

assessed as having been a product of the Industrial Revolution! Much to my relief, everything worked perfectly and only minor adjustments were needed to the undercarriage up-locks and the leg doors. The wheels fitted beautifully! Two problems did arise, though. One actuator for the radiator flaps leaked, and we had not catered for the flow rate induced by the rig. Hydraulic fluid cascaded out of the reservoir and all over the lovely, clean engine!

We planned to attach the propeller at the earliest opportunity, but several obstacles lay in the way. Firstly, the spinner had been badly distorted as a result of damage received in the 1950s. Consequently, it could not be made to fit the backplate, which had to be attached to the propeller before it could be slid on to the engine prop shaft. We had to eliminate the damage. The assembly found its way to Warton where its shape and correct profile, incor-

Nose view of substantially complete Gustav. *The supercharger intake can be seen on the left cowling.*

The electricians have been busy. The cover for the cannon is between the rudder pedals.

porating the exit port for the cannon ordnance, were restored. But our problems were not yet over, as we discovered that the propeller was still deficient of several roller bearings. Replacements were quickly found by John Rumbelow and, after Roger Slade manufactured the remaining parts needed, we were able to sleeve our beautiful propeller on to the front of an equally impressive engine. It was a lovely sight!

After months of bitter inactivity, we were suddenly inundated with work. There was always something — many things — to be done. We had no complaint about that, but we did wish that we could devote more time to the aircraft. An uninterrupted spell of a couple of weeks would have been perfect. There was nothing to be gained dreaming, however. As the electricians had made good progress in the cockpit (I could always tell by the piles of cable clippings) we badly needed to fit the instrument panel. Installing the instruments to the magnificent new panel presented a few unexpected problems, but they were mere 'hiccups' compared to designing, forming and fitting the various pipes which fed the pressure instruments. The completed panel was a Dixon work of art, enhanced by various metal labels made for us in Germany through the good offices of the ever-helpful Peter Nolte. I was able to reproduce various electrical circuit codes and instrument markings by referring to photographs of the original panel taken during the War and in 1961.

It began to dawn on us that little remained to be done before we would be in a position to begin engine-running! At the weekends our corner of the hangar seethed with activity as result. Apart from the four regulars, Ian Mason was able to lend a hand, although his new home near Bedford made

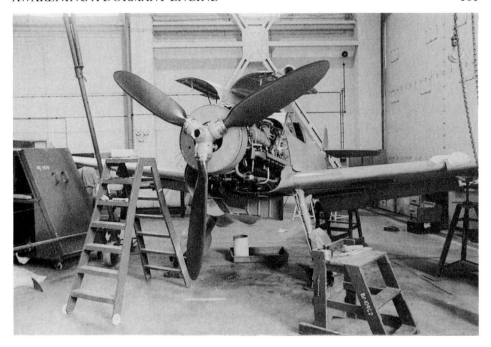

The beautiful green VDM propellor in place.

attendance difficult. Roger Slade beavered away putting the final touches to the engine installation assisted by Bob Kitchener. Close by a new face could usually be found. Chris Starr, a Tornado pilot from RAF Marham, had phoned me one evening expressing a keen interest in the 109 and asked to look it over. I was impressed by his knowledge, and over the course of a few visits he was soon deeply involved. His efforts, particularly with regard to the power plant, proved invaluable. I am not sure that Roger appreciated his

Fitting the prop, blades covered with polythene to avoid damage. On the left is Roger Slade, with Bob Kitchener in foreground.

How do you get fuel in this thing? Left to right: John Dixon, Paul Blackah and Roger Slade.

attention to 'his' engine, and I found myself constantly on the alert for possible friction. Thankfully, it rarely appeared.

The subdued excitement which affected us all served to concentrate our attention on the outstanding work required before attempting an engine start. The wiring in the cockpit at this time occupied the attention of no less than four electricians, including Eric Faithfull. I had no idea that so many existed! We lesser mortals assisted where we could.

One afternoon in May, the old bird was pushed outside the hangar doors into the sunshine to be fuelled. The Germans had a special fuel nozzle which was inserted into the filler neck and locked. Fuel was then pumped into the tank until a reverse flow was detected. Unfortunately we had no such device,

Sergeant Bill Dowie casts a critical eye over the rebuilt engine.

and the tank could not be filled car-fashion because of the attitude of the aircraft when standing on its wheels. Roger eventually designed and made a large funnel which, after a deal of trial and error, was found to work satisfactorily. But the first attempt was a disaster. The fuel went everywhere except into the tank! The situation was alleviated by use of the new funnel and by venting the top of the fuel tank. Once we had fuelled, we discovered several small leaks in the most awkward place (where else?) at the bottom of the tank. The only way to rectify the faults was to drain the tank, and this we had to do twice.

I decided, as we approached readiness, that Roger Slade should have the honour of being in the cockpit. He had been largely responsible for the latter stages of the engine rebuild and was a qualified pilot. However, reasoning that two pairs of eyes were more likely to spot anything untoward than one, I planned to be standing on the wing, looking over his shoulder.

Cockpit preparation prior to start is mainly common sense, but it may prove of interest if I run through an abbreviated checklist and start drill of the Daimler-Benz engine as installed in the Bf 109 G.

Attention is first paid to safety items. The undercarriage is checked in the down position by confirming that the appropriate push button is home and guarded by its spring clip. Above, the magneto switch should be off, the fuel cock closed and the battery switch off. Once settled, power may be selected by moving the battery switch on the right wall forward. Locks are removed and the flying controls checked for full and free movement. It is then easiest to work from the left side of the 'office'. Flap is wound up by use of the outboard of two large wheels. Winding the other wheel adjusts the tailplane position to a tail-heavy trim. Next, the fuel injector cut-off handle should be fully forward. Going up now, the propeller control switch is confirmed in the *Hand* (manual) position and the indicator on the instrument panel should show '12 o'clock'. Immediately above, the fuel cock must be placed fully forward and into a gated position marked 'P1 & 2', indicating that both fuel feeds and filters are ready for action. The throttle is set one-third

8 July 1990. The engine has just started. Roger Slade in cockpit.

A happy team after the first engine run. Left to right: Chris Starr, John Elcome, John Dixon, Roger Davies, Paul Blackah, the author, Bob Kitchener, Roger Slade and Graeme Snadden.

open next to a red line on the quadrant. Having checked two green lights on the undercarriage indicator, the magneto switch is moved to 'M1 & 2' (both magnetos). A quick look at the coolant temperature is followed by

The team presents a photograph of the Bf 109 to a delegation from the Jägerkreis, the German fighter pilots' association.

pushing the button on the instrument. The same needle then registers oil temperature. Below, sufficient fuel is indicated and oil and fuel pressure at zero. On the electrics panel, a selection of circuit breakers (CBs) — those needed for the task ahead — are made.

We are now ready to start the mighty Daimler-Benz. The engine, as fitted to the 109, is started by hand-cranking an inertial starter. (It can be adapted to start by electric motor, but the fighter was never thus-equipped, presumably the idea being to save weight.) So, a handle is fitted to a drive through a cut-out on the right engine cowling, and wound. This is when life becomes hectic for the cockpit occupant! First of all, several strokes of priming fuel are pumped into the boost pipes of the engine. When it seems that the 'handle-winder' is into his stride, pushing in the relevant CB starts the immersed fuel pump and pressure is seen to build on the left side of the dual pressure gauge. Book pressure reached, the CB is tripped. A couple more strokes of prime as the start handle is withdrawn, the clear signal given by the ground crew, a guard is lifted and the start switch pulled. In so doing, the ignition circuit is completed and the rapidly spinning flywheel is mechanically clutched. The engine fires, the start switch released, fuel and oil pressure checked and a good warm-up rpm set. All this happens in a few seconds — in theory!

Early in June we felt ready to go. I wanted everyone there for the event; they all deserved to witness the engine burst into life. One beautiful Sunday morning, 10639 was pushed out of the hangar, down a slight incline to a small area of tarmac close to airfield perimeter track. We manoeuvred the nose into wind, chocked the mainwheels and tied the tail down to several concrete weights. The moment had arrived for Roger to enter the cockpit. Once in, I climbed up behind, while on the right wing Paul Blackah readied himself to wind the handle — a new one made by Roger. (One day I hope we will find an original!) He soon felt the tremendous amount of effort required to cause the flywheel to move! The noise of the spinning mass reached a crescendo and Roger pulled the start switch.

The propeller moved only slightly under the influence of an audibly-slipping clutch. So much for my hopes of a fairy-tale first-time start! Not daunted, we repeated the rigmarole, with the self-same result. After several more attempts it was time to call a halt and diagnose our problem. I had seen lots of film of 109s being started during the War, and it bothered me that our prop scarcely moved. However, we decided to look at other possible faults. John noticed that there had been no noticeable smell of fuel from the exhaust ports. This prompted examination of the fuel cocks and filters under the floor. Removing some spark plugs confirmed no smell of fuel from the cylinders. Some more winding by Bob Kitchener offered little hope of an imminent start, and a slightly disillusioned team pushed our old fighter back home. The return journey to the hangar was not so easily accomplished because of the slight uphill slope. Although small, the Bf 109 is really quite heavy! It was the first of many such energetic outings attempting to breathe life into the engine.

The following weekend, we gathered again and completed further start cycles. We got excited when the propellor moved one blade — which is a fair

Yet another engine run! Paul Blackah begins to wind for Roger Slade.

indication of the total lack of success. Our next investigation examined the spark plugs, some being replaced and others having their electrode gaps adjusted. At John's suggestion we moved the prop by hand until the piston of the No.1 cylinder was on the compression stroke. His idea was to pour a small quantity of neat fuel into the pot and the following two in the firing sequence, screw in the plugs and try a start. There was a noticeable difference, a puff of smoke emanating from one exhaust stub, proof of ignition at least. But the majestic propeller hardly flinched. Bob made adjustments to the clutch cable, but again, no indication of life.

A week later, more of the same. This time we changed batteries regularly

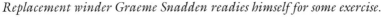

Replacement winder Graeme Snadden readies himself for some exercise.

The first public view of the Messerschmitt. The Open Day at Benson, July 1989.

and dabbled with a prime fuel of higher octane. Bob deduced that the starter clutch was slipping excessively as a result of the flywheel being overcranked. His experiments with a slower rpm certainly had effect, the propeller turning through one complete revolution on occasion, sufficient, in my opinion, for an engine start to have resulted. The situation was beginning to depress morale to the extent that I decided we should continue our investigations mid-week whether all the team could attend or not. But several full work days later, we were no further forward.

Their first sight of the 109 in 45 years. Left: Bill Dowie and right: Doug Gough, both formerly of No. 1426 Flight.

Sunday 8 July dawned a beautiful day, and we assembled in the hangar ready for another session. Most of the earlier enthusiasm had evaporated, but we were all eager to get started. Several hours later, having exhausted a succession of 'handle winders', further investigation was called for. We knew we had a spark at the plugs, that fuel from the tank was available and we were certainly using prime fuel by the tankful. So why wouldn't the d***** thing fire? The DB 605 engine is large, and we wondered whether the prime pump was 'man-enough' to propel a charge of fuel all the way to its front end. We soon discovered, to our consternation, that the answer was no. The priming tank was removed and dismantled on the bench. It had always seemed to us to be an overly 'fussy' device. Perhaps this had been the reason that Messerschmitt replaced it with a more simple pump on the later Bf 109 G. Laid out in its component parts, we still could not fathom the reason for its failure. Finally, I suggested reversing a small shuttle valve inserted in the delivery port, although, I confess, it seemed to make little sense so to do.

Assembled once more and topped up with fuel, we were amazed to see a charge of fuel shoot a great distance across the hangar floor! Enthusiasm rediscovered, the tank was put back and Bob wound the start handle. This time the propeller 'kicked' — a sure sign of life, at long last. But further cycles were despairingly lifeless. It was getting late, and I decided to call it a day. I was overruled by a chorus of protest. 'O.K.,' I said, 'but only one more go.' Flywheel whining, Roger pulled the start switch, and, blow me, the engine started! It took me a couple of seconds to recover from shock and survey the cockpit. All seemed in order, except that the fuel pressure was

Investigation of the fuel system at Benson. Bob Kitchener kneels behind the wheel with Roger Davies while John Dixon inspects the top of the fuel tank.

'off the clock'. (It transpired that we had the wrong dual pressure gauge. I found the correct one some weeks later — in Belgium.)

As planned weeks before, Roger stopped the engine after a few minutes to allow a check for obvious leaks. It was a real pleasure for me to see the smiles on the faces of my colleagues and join in the round of congratulations. It was a moment of supreme elation. Interestingly, there were a number of bystanders who were certainly not there before the engine started. I discovered that they all came from station married quarters which were downwind of the aircraft that day, but over half a mile away. Such is the distinctive growl of a Daimler-Benz engine! Roger restarted the warm engine a few minutes later to satisfy us all that the first time had been no fluke, and we put the aircraft to bed. It was only when I was driving homeward with Graeme that I realized that my left ankle was tingling. While standing on the wing, my foot had been only a short distance behind the left exhaust bank. I had not noticed the heat! It was some days before the limb returned to normal.

14

A second maiden flight

The following weekend was devoted to more of the same. Whilst we could start the engine, the method we employed was not as described in the *Handbuch*. The throttle position was far removed from one-third open, and many more strikes of prime were needed. The first we adjusted, but Roger felt that it was a peculiarity of our engine that it liked a rich fuel mixture. I could not agree and began looking for a substitute priming fuel. In the meantime, at least, our non-standard actions provoked an engine start. There were days, though, when many attempts were needed and I was convinced that a large part of the responsibility lay with the starter clutch.

Bob Kitchener had established himself as the most successful winder, and provided he had any energy, we expected him to be on the right wing. His system worked, after a fashion, but it was far from satisfactory. Provided the rate of crank was kept just below one revolution per second, the propeller could be made to turn a sufficient distance for the engine to fire. Faster than that, and the clutch protested vigorously as it slipped. The *Handbuch*, on the other hand, advised a crank rotation of over 90 revolutions per minute as optimum. Designed to slip, of course, it was apparent to me that it was not behaving correctly. Adjusting the operating cable had no significant effect and, as removing the starter entailed first removing the engine, we elected to endure the situation for a while. (I was to curse that decision some months later. In front of a large audience, including representatives of the Ministry of Defence, it took me several hours to start the engine!)

My next concern in the test programme was to optimize the engine/propeller interrelation. As I have already described, the Bf 109 system relies totally on the position of the throttle and the power demanded of the engine, the propeller pitch automatically adjusting to suit. With that in mind, it follows that, having set a power figure on the boost gauge, the propeller will be made to turn at a calculated rpm. During our initial power runs we found that very little throttle was needed to achieve high propeller rotational speeds. Baffled by this, I asked Roger to alter the boost setting. It was a bit like feeling in the dark really, fiddling with an adjustment screw, but it did have a small effect on performance. Chris, however, put forward a suggestion which was pure common sense. If little power was needed to rotate the propeller quickly, it could only be that the pitch had been incorrectly set. A more coarse angle on each blade would increase air resistance to the movement of the airscrew, requiring more power to achieve datum rpm. The reasoning was sound, but the propeller had been delivered with each blade set in clamps to hold them at the correct angle — or we assumed that they were.

A warning bell sounded in my cranium, and a rule which had been instilled during my Air Force days sprang into view: 'Never assume — check!' Paul used a clinometer to check each blade angle. Two were almost correct, but the third had been inexplicably set some seven degrees too fine! The findings were double checked before the propeller was slid off the drive shaft and each blade correctly set. The next engine run proved remarkably close to book figures. 2,100 rpm, for example, was attained by setting just under one atmosphere on the boost gauge.

Over the course of successive weekends we slowly refined engine operation — Roger and I taking turns in the cockpit. We had several ever-present faults during that period. The oil function of the double-temperature gauge could only be made to operate intermittently. As it had been overhauled, we thought the temperature sensor on the engine to be the culprit. It was removed and taken to Bristol for examination. Exasperatingly, it was found to be serviceable. In fact the gauge itself was found wanting. Although it bore the correct part number, its construction was too delicate to cope with Bf 109 requirements. I acquired another some time later, which, though bearing the self-same part number, was of different manufacture and much more robust in construction.

I was puzzled by the behaviour of the radiator flaps during power runs. Selected manually, they operated correctly. However, I could get no response

In the midst of a power run. Radiator flaps are fully open and the exhaust stain is already coating the wing-root.

from them when set to automatic operation. Despite coolant temperature hovering near the maximum permissible, the flaps remained obstinately shut. The cause was a slight embarrassment to us all; we had inadvertently transposed the hydraulic hoses feeding the thermostat control. Thus, although the device was functioning, hydraulic pressure had been directed to the wrong side of the flap operating ram. Instead of opening the flaps, it had consistently closed them! A silly little mistake which one of our number was not allowed to forget for some weeks!

Our progress with the engine was necessarily slow. We trod each step carefully for fear of damage resulting. Few Daimler-Benz engines have been run in recent years and advice was not always available. Messerschmitt-Bölkow-Blohm, for example, were having problems with the DB 605 D engine installed in their Hispano HA1112/Me 109 G-6 hybrid. It had (reportedly) seized in flight and a total rebuild had been necessary, no mean task. The Finnish Air Force Museum had last run the engine of their machine many years before and were unlikely to be of help, had we needed it. We were very much on our own.

The motor sounded rough, particularly at low speeds, and it seemed that some cylinders were 'missing' on the ignition stroke. Even at higher speeds, an occasional 'crack' indicated a possible fault in a spark plug or its wiring. In the cockpit, an unacceptably large drop in rpm when checking each magneto confirmed the problem. Weekend after frustrating weekend we removed spark plugs and either replaced them or re-gapped them. There was an immediate improvement each time, but it was fleeting, the mag-drop recurring after a couple of engine runs. It was to be some months before we discovered the true nature of the fault.

When they could, meantime, John and Paul continued their work on the panels supplied by the Finns. The most complex fitted around the leading edge of each wing-root and joined underneath. Handmade by Valmet, both were far from being a perfect fit. As Benson did not possess the facilities needed to reshape them, for a second time I approached Abingdon with official backing. In remarkably short time, the alterations we required were completed and the panels returned to us. If only we had had similar co-operation over preceding years when we badly needed specialist assistance. Several days were spent trimming each panel in turn, following which a veritable jigsaw of small alloy parts and plywood stripping, supplied by Tikkakoski, were affixed. Neither craftsman enjoyed preparing these parts. Their three-dimensional configuration rendered them difficult to manage. But once in place, they looked satisfyingly correct.

The last panels to be tackled were intended to fair the top surface of each mainplane to the fuselage sides. Beautifully made by Valmet once again, they, too, needed adjustment, but the lads were able to handle this work themselves. The method of attaching these long fairings is interesting. A cable attached to its lower face is hooked at two points by turnbuckles attached to the fuselage. With the panel lying in its approximate position, the cable and turnbuckles can only be reached through a small access hatch underneath. Turning knurled wheels shortens the length of each turnbuckle, reducing their reach and drawing the cable and the fairing downwards to form a snug fit, another example of intelligent designing making life easier for the mechanic.

I had promised to keep MoD informed of my progress. The relevant branch was involved in drawing up a contract of loan, and I had insisted that I was consulted on any portion which affected myself and the team. A few minor disagreements arose, and the offending paragraphs suitably amended. Strange to relate, I was also being hounded to provide information which would lead to the Historic Aircraft Committee parting with a great deal of money! I had casually commented over a year previously that I would have preferred to fit three composite wooden blades to the propeller. Had it contacted the ground at any time, the wooden blades would have shattered, but there would have been no shock damage transmitted to the engine. It was a passing remark, so I was amazed that HAC had decided to provide funding. It was even more astonishing when the finance required is considered. No paltry figure, each blade was costed at £3,000! On reflection, I was inclined to feel that the same money would have been better spent on one of the more neglected aircraft, but I could not dissuade Colin Cheesman. An order was placed with Hoffmann Propellerwerke in Bavaria.

As the testing phase developed, Service minds turned to the subject of planning an official roll-out ceremony. At one of the continuing meetings at Benson, I concluded that we could feel reasonably certain that the aircraft would be ready in May. For the first time I had agreed a target date, and I had a nagging suspicion that the time allowed would prove inadequate. It always is, and this was to prove no exception. Parkinson's Law!

Group Captain Reg Hallam, our nominated test pilot, began spending time in the cockpit and asking pertinent questions. Reg was then Officer Commanding Experimental Flying at the Royal Aircraft Establishment at Farnborough. A graduate of the Empire Test Pilots School, he was in current flying practice on the Spitfire, Mustang and Hispano Buchon. This last, while derived from the Bf 109 G, was powered by a Rolls-Royce Merlin engine. The cockpit had been redesigned and reflects the hybrid nature of the machine. It is radically different, therefore, from the Bf 109. Added to this, everything in 10639 is labelled in German. A lot of explanation was needed to educate our pilot!

Towards the end of January, I began calling Reg at home to keep him in the picture. I hoped that we would be ready for our first flight in February. A Permit to Test had been granted by the Civil Aviation Authority, and the aircraft allocated the distinctive identity G-USTV. It was as close to Gustav as the British system would allow! So legally we were ready, and the aircraft, although not perfect, was almost serviceable. The weather kept us indoors for some time. One short excursion to the airfield allowed me to check the generator which had been refusing to function. It had been overhauled twice, the last time by a local expert who had demonstrated a healthy output on his test-bench. On that day, I found that it was working perfectly when driven by the engine, and our last major hurdle had been taken. Rather than push the 109 up the slope to the hangar (the effort always left us breathless), I suggested to Reg that he taxi it. The rest of us scampered around like demented chickens as the aircraft moved from the apron, the first time it had moved under its own power for 45 years!

Further wild and wet weather forced a postponement of our maiden flight. Benson had been preparing a grass strip for us. Unused for years, it received several rollings and the grass was kept short, but it was scarcely ideal. Lying alongside the airfield's shorter runway, the surface was uneven and featured a pronounced slope. Worse, and unknown to me until a few days before our flight, two trenches had been dug across to lay cables for a nearby radar caravan. Although adequately filled, the surface had been replaced with squares of turf and these had not had sufficient time to knit together. I inspected them with Reg and expressed my worry that the high wheel-loading of the fighter would plough through the turves. I suggested that they were removed and a temporary surface of ash or similar laid, but Benson decided instead to continue rolling.

A last vital task was performed for us by Abingdon staff — checking the weight and balance of the aeroplane. Its weight proved to be close to figures quoted in various publications but, to my astonishment, it was tail-heavy! As we had not fitted the bulky radio equipment or compressed air bottles in the rear fuselage, I had expected the reverse.

The weather eventually relented, if temporarily, and on Sunday 17 March we all gathered in the hangar. It was an overcast day with a moderate breeze blowing. The grass was wet, and the strip had not really improved over the intervening days since I had last seen it. Reg decided that conditions were suitable, if hardly perfect. There was a slight hitch, though. John Dixon had taken time to examine the weight and balance data and discovered an elementary error in the calculations. Correcting this confirmed my earlier suspicion that the aeroplane tended towards nose-heaviness. We had to establish the exact position of the centre of gravity, and pieces of paper adorned with obscure calculations passed to and fro until a unanimous conclusion was reached that it was within the envelope, but only just. Reg was satisfied. The

Reg Hallam settles in for the maiden flight. He is strapped in by Paul while Roger watches anxiously.

team finished the preparation of the aircraft while the pilot gathered his thoughts.

After the usual reluctant start performance, chocks were removed and the 109 trundled along the perimeter track onto the main runway followed by myself, Roger and Graeme in his car. Reg next turned onto the short runway which rather surprised me, as we had previously discussed back-tracking the grass in order that I could inspect the effect the wheels of the Messerschmitt had on the resurfaced trenches. Run-up completed satisfactorily, I walked with the aircraft as Reg lined it up on the grass, noting that in the right turn, the mainwheel was fully braked but slid easily on the surface.

A nervous last-second inspection, and I gave the pilot the 'thumbs up'. The Daimler-Benz growled smoothly and 10639 bounded forward on its first, post-restoration take-off. It was 12.57 Greenwich Mean Time. I cannot describe the pride I felt at that moment, for it was heavily tinged with anxiety. Nearly 20 years of work had been lavished on the little fighter. The tail rose after a roll of but a few yards and I noticed Reg applying full right rudder, countering the torque effect of the large propellor. Nevertheless the 109 slowly veered off-line to the left. A scant few seconds later it traversed the first of the trenches and my fears became reality. The wheels dug deeply into the new turf and clawed some of it free, and the gentle swing from the strip centre-line was aggravated. Shortly after, Graeme noticed a substantial amount of grass being propelled towards us. I confess that I turned away briefly as I was convinced that an accident was imminent. The 109 (and Reg!) had different ideas, and it was launched into the sky after encountering a small hillock in the grass.

With great relief, I watched as the only Bf 109 to fly for 38 years accelerated away from the field, undercarriage slowly retracting. Reg remained clear of cloud and within sight of Benson while he explored the handling characteristics of his new bird. I had a portable radio by my side in order that I

The take-off. The aircraft is veering left having hit and lifted some loose turf, and the rudder is hard right.

The author about to tell Reg that his undercarriage is down and locked.

The team waits anxiously for the 109 to return. Left to right: Roger Slade, Dennis Rolph (a keen supporter), the author, John Dixon, Graeme, Paul and a helpful glider pilot.

Reg makes a low pass to check the landing strip.

could communicate with him, and I well remember his first call. 'Russ, it goes like a train!' Evident to all of us watching admiringly, but still anxiously, was a trail of smoke emanating from the engine. Indeed, the location of the aircraft was most easily discovered on the end of the plume. It worried Roger and me for a few seconds, but the reason was obvious. Over the left group of ejector stubs is fitted a deflector shield to guide the exhaust away from the supercharger intake. Thus directed, the trail appears remarkably dense, whilst the right exhaust is free to disperse.

Several minutes later Reg declared that he had problems with the undercarriage! A cold chill ran down my spine. I could imagine the ramifications of a wheels-up landing following the very first flight! After exploring the slow speed feel of the 109, the wheels had been selected up. Re-selected down, they refused to extend for some considerable time, during which I racked my brains for possible causes. Reg returned to the overhead and I was relieved to see both legs down. But then I learned that power had (apparently) been lost to the engine instruments! I had been joined by Paul Blackah and John

Undercarriage indication restored, Reg Hallam approaches touchdown at the end of the maiden flight.

Dixon when Reg announced that he had also 'lost' the undercarriage indicator and compass. With remarkable speed, Paul asked me to suggest checking a circuit breaker; he realized that all the circuits lost were fed via one CB. It had tripped. Paul received a heartfelt pat on the back! Reg planned to perform a practice approach and overshoot prior to landing, and during the low pass, I took the opportunity to examine the undercarriage through binoculars. They appeared 'down and locked' and the aircraft was flown around the circuit for its landing. The throttle was closed and the 109 floated some distance into the strip before touching down close to the trench which had been the cause of my first 'heart-attack' and rolled down the pronounced slope, coming to a standstill before meeting the metalled main runway. The time was 29 minutes past 1.

We were all very relieved, and tore around the airfield in our respective vehicles. A large number of people had gathered to watch our baby return to the ramp. As he opened the cockpit canopy Reg received well-deserved applause. But his — and our — moment of triumph was marred somewhat by the discovery that the propeller tips had been bent during the take-off run, much as Graeme had suspected. Reg had been unaware that it had happened, of course, the propeller having only 'feathered' some soggy grass for a few yards. Despite the damage, the propeller had still been able to haul the 109 through the sky at startling speed! Less serious, hydraulic fluid was pouring from every seam of belly panelling.

Back in the hangar, we shared a well-deserved bottle of champagne. It had been an eventful 32 minutes and an exhausting day. Reg was extremely upset by the damage to the airscrew, and he would not be consoled despite my assurance that the blades could be repaired. His report on the flight can be found in Appendix C, but I must add my own observations. Reg had intend-

Touchdown — the end of an eventful first flight.

One bent propeller tip. This was by far the worst of the three.

Aerial view of take-off path, aircraft travelling from top to bottom of picture. Skid marks indicate the right wheel is fully locked while incisions in the grass show where the propeller struck the ground. Top left may be seen touchdown point, and divergence of take-off track is evident by comparison.

The relaid turf showing just how deeply the wheels penetrated.

End of a climactic day. Left to right: Roger Slade, Graeme Snadden, Bob Kitchener, the author, Ian Mason, Reg Hallam, Paul Blackah, John Elcome, John Dixon and Chris Starr.

ed lifting the tail on the initial part of the take-off run much more slowly than he did. However, the engine response to throttle movement is rapid and it took him a little by surprise. The tail rose quickly, therefore, aided by the forward position of the centre of gravity and he found some difficulty in controlling pitch with elevator. Shortly after, the wheels dug into the trench, aggravating the attitude instability through a momentary deceleration, and resulting in a remarkable impression of a powerful lawnmower (according to Graeme) as grass was chewed by the prop blade tips.

Another facet of the take-off was directly attributable to the trench. The Bf 109 always tended to swing left because of the direction of rotation of the large propeller, but as it encountered the newly-laid surface, the left wheel dug in much more deeply than its brother, greatly aggravating the swing off line. Once such a swing has started, it is difficult to counteract, and examination of the grass afterwards revealed that Reg had applied full right wheel-brake, and the wheel had locked over a considerable distance, but even this, and full right rudder, proved ineffective.

In his report Reg comments on the length of the landing run. Of course it was his first in the Bf 109, but in fairness to the aeroplane, it should be borne in mind that the landing was downhill and on wet grass. Also, but unbeknown to Reg when he wrote his report, the wind had changed direction before his landing and a video of the event points to a slight tailwind. All these factors were scarcely conducive to a short landing run!

That first momentous flight was Reg's last. He retired from the Royal Air Force prematurely to take up an appointment in the Middle East. It was a shock and great disappointment to myself and my team. We had all looked forward to working with him. It was his style to keep everyone 'in the loop', and it was much appreciated. His skill in salvaging the take-off which, frankly, I found terrifying, was of the first order. Above all, he is a remarkably brave man.

15

Recovery and roll-out

A drenalin levels were maintained for several days, but it did not stop us investigating our known faults and giving 10639 a good examination. The cause of the loss of hydraulic fluid was found immediately after the flight. A union had not been adequately tightened on the pump — an oversight on our part. Fortunately, although the result was messy, the loss of fluid was not as dramatic as it had at first appeared. The strange behaviour of the undercarriage, though, was hard to fathom. At first I thought that it may have been because of the progressive loss of fluid, but such an explanation did not fully fit the bill. In fact, all the worry had been down to a small mal-adjustment within the selector valve. Once found, it was cured in minutes.

And the circuit breaker? It had tripped purely as a result of vibration and was rectified the following day. But the big imponderable was the damaged propeller. All three blades had been bent, of course; the worst over a length of some five inches. The question was whether the damage could be repaired without recourse to heat treating. If not, then I should have had to find in-depth information on the alloy material before any rectification could be attempted.

We contacted Hoffmann, but the response was far from encouraging. They thought that heat-treatment was, indeed, necessary and the company could offer no such help. I was not convinced. Fortunately, when the pro-peller lay at Lostock years before, and I searched for the information which BAe demanded, the Finnish Air Force sent me several pages from a repair manual. These were evidently Finnish translations of the German VDM manual and I managed to persuade the Finnish Embassy to provide an English version! They contained several explicit graphs indicating the extent to which blades could be bent before heat treatment was required. Referring to these, I had to disagree with the initial assessment by Hoffmann. Their reaction was immediate. The information was faxed to Rosenheim and, once more, the German propeller specialist came to our rescue by agreeing to repair the damage.

I was faced with the reality that there were less than six weeks remaining before the arranged roll-out ceremony on 2 May. I knew it had been a mistake to agree to a date! It was imperative that the prop found its way to Germany with all due haste. Colin Cheesman explored all the possibilities. At first, it was planned to fly it over by Hercules transport aircraft — an ideal solution, but it proved nought but a dream. Permission was granted to task an Andover from Benson, but the diameter of the airscrew was too great for it to be inserted in the much smaller aircraft, and I could not risk dismantling it to suit. Thus we gravitated towards the slowest possible mode of transport, a scheduled truck from Stafford to RAF Germany. Once collected, it took several days to arrive at its destination.

Despite assurances that the restored unit would return by the middle of April, alternate plans had to be made. The roll-out could not be postponed. Even if it did not fly, the 109 had to be on parade, preferably with a propeller! Provision was made for the removal of one from the Messerschmitt Bf 110 G-4 at Hendon, should it have proved necessary.

The accident had postponed our test programme which had been set at a total of six flights. Even if the prop did return when hoped, it was unlikely that we could complete it before the big day. In my own mind, I felt that preparing the aircraft to allow our new chief pilot, Air Vice-Marshal John Allison, a couple of flights would be a more realistic target. A flight on the day would have to be classed as part of the programme.

While we waited impatiently, the engine was inspected for damage. The main worry was that, in 'mowing the lawn', the heavy propeller may have damaged the hollow drive shaft. It had been unaffected, much to our relief. A few days later, news percolated through to me that we were unlikely to see our repaired airscrew until much later in the month. The test programme became less important than making the aircraft presentable, and I negotiated with the hangar to provide space suitable to allow me to paint the 109. (Our sand and light blue spray dust would not have been popular all over a predominately white and red Andover!) One weekend was allowed, and half the

10639 with basic camouflage colours applied and wing Balkenkreuz *being masked.*

John Dixon sprays a stencil on the nose cowl.

large building vacated. A few days before, John Elcome, Paul Blackah and myself began to mask off. The paint team under the guidance of Corporal Bob Nash joined in with gusto and the preparation took only a few hours as a result.

On the Saturday morning we layed acres of brown paper on the hangar floor, and the scene was set. Bob and one of his men began bathing the 109

The yellow lower cowling displays its identification value to the observer on the ground.

in an overall coat of light blue (RLM 78). We had become used to seeing it in a shade of grey-green which had become grubby. Under the glare of the hangar lighting, we watched the paint begin to dry and the clouds of spray dust subside. It looked absolutely beautiful!

Next day, some masking was removed and yet more applied and we were ready for a coat of sand coloured paint (RLM 79). It was vital this time that Bob was aware of where to point his spray gun. The paint finish, after all, was a vital part of accurate restoration. I spent some time showing him photographs and going around the aeroplane lightly marking the fuselage to indicate the demarcation line below which his spray gun dare not trespass — on pain of instant death! I was never more than a few feet away from him as he followed my instructions, ready to stop him should he seem to be about to make an error. No interruption was necessary and, again, we were able to admire the aeroplane in yet another new guise.

The following evening, when work had ceased in the hangar, we applied yellow to the lower engine cowling. the first step in a concerted week-long effort to finish the paintwork. It had taken some years of research, incidentally, before I realized that the aircraft bore this colour — as did most Bf 109 F and Gs in the desert, I discovered. I have found no official explanation for it, but it is reasonable to assume that it was to aid identification from the ground. The 'theatre markings' followed, white wing tips, spinner and broad band around the rear fuselage. I had all the information I needed on the size and position of the national markings. The upper wing and fuselage *Balkankreuz,* though, were predictably hard to apply correctly because of the varying contours of the surfaces on which they were applied. Similarly, the *Staffel* number proved a real headache. Examination of photographs revealed that the figure was not regular in shape, and I feel that the painter who applied it did not have the correct stencil and created it using a combination of part stencil and part masking tape. 10639, for some reason, did not carry a full set of stencilled instruction markings. As my aim was to restore the aircraft to its original state, those markings stipulated in the German manual, but omitted

The right fuselage complete with all markings. The irregularity of the black '6' is deliberate.

Pushed out into the daylight for the photographers — a pristine Bf 109 G.

on the aeroplane, were not applied. On the leading edge of each wing was a reminder to groundcrews that the Bf 109 G-2 could only carry the early, small wheel in the mainplane wheelwell. Unfortunately, in April I had only one vague view of this marking. Obviously applied freehand and probably in red, it appeared to read '*Großtmaß 160x660*' — the maximum size of the tyre/wheel. It took me a few hours, but I was happy with the result. (Some months later, I was to receive several more photographs which clearly showed that I had guessed wrongly. The inscription actually reads '*Achtung! Großtmaß 160x669*'. I did not even manage the correct colour! '*Achtung!*' was in red, but the rest was in black. I corrected the error at the first opportunity.)

The aeroplane had not borne a unit marking, probably because there had been no time to apply it. I conferred with my colleagues as to whether we should apply the crest, a wolf's head on a heraldic shield, as a tribute to the unit. We all felt that it should be applied, and I decided to apply it to one cowling only. I hope that the purist will forgive me this one transgression!

Over the preceding years, when asked, I stated my resolve to paint the aircraft precisely as it had been found. The usual retort was that it was too dull, and I was encouraged to give it a more exciting identity. I wish the sceptics could have been in the hangar at the end of our labours. The Bf 109 was very smart, even beautiful! The only conceivable criticism could have been that it was too pretty, but it had yet to fly in its new coat and I was sure that exhaust and escaping fluids would eventually dull its plumage. I realized it was bound to happen, but it was not a prospect I particularly relished.

Paul called me at home a few days later to announce the arrival of the propeller. It had been a startling performance by Hoffmann, who had done their job in only a few working days; the rest of the time being spent in transit. The tone of his voice, however, indicated that more news was to follow. 'They've painted the tips bright yellow!' As I was unable to return to Benson for some days, I asked him to contact the painters and arrange a hasty respray. In the meantime, the propeller was installed, and the following day a further coat of *schwarzgrun* was applied. When I next entered the hangar, the original appearance of 10639 had been fully reproduced. It gave me great pleasure admiring it from every angle. I could still remember its shambolic state some 19 years before (was it really so long ago?).

It was just possible that we could resume the test programme before the roll-out which was less than one week away. John Allison was free and, unexpectedly, Reg Hallam was available to coach and guide. On Sunday 28 April (four days to go), I entered the cockpit to perform a brief engine run. I wanted to make sure everything was serviceable before handing over to our new pilot.

Several hours and several tanks of priming fuel later, the engine showed no sign of life. Once again the prime pump was the culprit; a seal had torn. When replaced, the Daimler-Benz burst into life, but I was immediately conscious that all was not well. Even at normal warm-up rpm, everything in the cockpit was vibrating. Minimum temperatures attained, I performed a magcheck, and the result was dreadful. Selecting either magneto incurred a loss of over 200 rpm.

Clearly something was badly wrong. Chris Starr was hopping about excitedly outside gesticulating towards the exhaust stubs. Slowed to idle, the engine sounded much like a steam engine — off song. Following shut down, I learned that occasional sheets of flame had been seen coming from one of the exhaust ejectors on the right side, towards the front of the engine. Once it had cooled down, the stub was removed and our problem was evident; an exhaust valve had burned through. It was a shattering, despairing moment for us all. There was nothing for it but to remove the engine and strip it down. But this would have to wait until after the roll-out.

Thursday 2 May dawned a grey overcast day with a biting wind. It was just as well, perhaps, that the proposed flying display for the throng of invited guests could not have taken place. The Messerschmitt lay at one end of a hangar, placed there the previous day in front of a frightening number of rows of seats. Mid-morning, to please the photographers present, the hangar doors were opened to improve light levels. It may have pleased them, but the Station Commander, Group Captain Greenway, on arrival in advance of the guests, was extremely angry. The building had only a minimum of heating, and although tolerable, the temperature had not climbed to a comfortable level. And then some idiot opened the door and allowed the gale in! I was glad that it had nothing to do with me.

As there was little time before the speeches were scheduled to begin, I was only able to meet and chat briefly to a few of the principal guests. I renewed my acquaintance with Bobby Gibbes (still in his usual sparkling form) and his wife Jeannie. I was also introduced to the captor of the 109, Wing Commander Ken McRae, a tall Scot who had, somewhere along the way, developed a strange accent. (He has been resident in Australia since the War!) We all took our places: myself, John Dixon and Paul Blackah alongside

Bobby Gibbes poses happily in front of 'his' Gustav.

John Allison, all facing the audience. I was a little taken aback by the arrangements, as I had expected the rest of the team to have been with me.

The ceremony began with a presentation by the Station Commander on behalf of Benson, and he was followed by Michael Fopp whose speech took the form of a historical account of the Bf 109 and a most complimentary résumé of my project. With apologies extended to all present that the aircraft was unable to fly that day, a short video was shown of the maiden flight. The lighting in the hangar made sight of the screen difficult, I am glad to say!

Heinz Langer, Lüdemann's nephew, chats to Bobby Gibbes.

The new Air Member for Supply and Organization added his congratulations, after which I was to receive an unexpected surprise. I was presented with a Revi C/12D gunsight, flown over from Germany in a Dornier Do 28 light transport that morning! Some months before, I had received a visit from Günter Leonhardt who was rebuilding a Bf 109 G which had been recovered from the Mediterranean, and I was able to offer him some help. Unable to attend the roll-out personally, he had persuaded a Luftwaffe General to deliver his gift to me!

The time had come to unveil the reason for the gathering, and I was gratified by the level of applause from the guests when they had their first sight of 'Black 6'. I felt like applauding too, if only to revive my hands! The ceremony, though not long, had allowed the hangar temperature to penetrate to the bones. The doors were opened once again and the team and I pushed the aeroplane to a spot suitable for photography and for the guests to examine it. It was a painful move for me! I was pushing the right wing leading edge, close to the undercarriage when the aircraft slowed. Assuming we had reached the appointed spot, I got behind the leg to help bring it to a halt. It seemed to stop, and I straightened up. Before I could get out of the way, the 109 resumed its progress and the wheel rolled over my foot. 'Was that painful?', asked John Allison, rhetorically. Fortunately, my eyes had been watering for some time anyway because of the wind!

Later, we retired to the Officers' Mess for lunch. But first, in the anteroom, Air Chief Marshal Sir Patrick Hine said a few words and introduced me to Oskar Friedrich, the Head of Messerschmitt-Bölkow-Blohm, who pre-

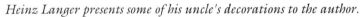

Heinz Langer presents some of his uncle's decorations to the author.

Distinguished visitors. Left to right: John Dixon, Paul Blackah, Ken McRae, Bobby Gibbes, Doug Gough, Ian Mason, the author and John Elcome on roll-out day.

sented me with a pair of replacement valves for our engine. It was a lovely gesture. There followed yet another surprise. Heinz Langer, the nephew of the last Luftwaffe pilot to fly 'Black 6' in combat, took the floor. His grasp of English was disgustingly good, and he gave a brief account of his uncle's career before presenting me with two of his medals and a wrist compass inadvertently left behind when his *Staffel* made its way to North Africa. I intend these to be displayed in front of the aircraft upon its retirement.

I then took the opportunity to talk to the personalities who had figured so prominently in the project, including Jack Bruce, now in retirement but kept busy following his abiding interest in early aviation. Peter Nolte, whose contributions over many years had been so vital, had made the long journey from his new home in western Wales. And Andy Stewart, whose tireless efforts in arranging the success of my work continue to this day. We all enjoyed an excellent lunch, and the event drew to a close as the guests left to make the journey home. It had been an exhausting day. I do not think that I have ever talked to so many people, and yet I had not managed to find time for a number of the guests. They had all enjoyed the day, despite the inclement weather, but I regretted not having had more opportunity to thank them all for their contributions to the project.

16

The search for history

aving painted 10639 as *'Schwarze 6'*, it follows that I should offer a few words to explain the discovery of its identity! It is a complex tale which required years of searching by Ian Mason and myself. Latterly, Andy Stewart became engrossed in the subject, yet we still do not have all the answers!

We must return to 1972 and the myth that had been built around the aeroplane: that it had been captured in Sicily in 1943. It took me some time to appreciate that before the fall of the island the Luftwaffe had received the more heavily armed Bf 109 G-6. Several of the mark were captured in good order, so why would the Allies bother with an obsolescent Bf 109 G-2? It did not make much sense. I reasoned that the search for the truth had to be focused on the earlier campaign in North Africa.

The breakthrough came in November 1976. In the readers' letters pages of *Air International* magazine, Group Captain Keith Isaacs, RAAF, submitted a beautiful picture of a captured 109 bearing the codes CV-V and about to take off from a desert strip. Misidentified as a Bf 109 F, it was unmistak-

The signal which caused great excitement in the hushed portals of the Public Record Office. Unusually, and thankfully, the scribe correctly identified the subject aircraft.

```
        M.E. EAI/580 of 10.1.43. Me 109 G.

     The chief differences between the Me.109F-4 and the
Me.109G-2, No.10639 being tested at Lydda, are noted below:
1.   The boundary layer by pass ducts normally found on the
Me.109F coolant radiator system are omitted on the Me.109G-2.
2.   A shut off valve is provided in the inlet and outlet
pipes of each radiator, each pair being operated by a ring
attached to a bowden cable near the floor on either side of
the cockpit, painted green and marked "Kuhlerabschaltung".
```

Top: *A composite view of PG+QV, W.Nr. 10652, a close relative of Black 6, at München-Riem in October 1942. This machine became Black 2. In foreground Günter Behling and behind a selection of* Gustavs *flown by III/JG77 from Russia.*

Above: *Black 6 as found by 3 Squadron on 13 November 1942.*

Below: *The Australians inspect their find at Gambut, 13 November 1942.*

ably a Bf 109 G-2. His letter revealed that it had been flown by Squadron Leader Bobby Gibbes at Gambut, Cyrenaica, soon after the Battle of El Alamein.

There was no guarantee that this was our aeroplane, but it was the first tangible lead I had. Keith put me in touch with Bobby, and there began a regular correspondence between Australia and England whilst I found out as much as I could. Fortunately, he had kept a diary and we were able to reconstruct the capture of the aircraft over the following months. He had taken to flying mock combat sorties against his unit's P-40 fighters and soon found that the obvious superiority of the German fighter was in danger of demoralizing his men! The prize was escorted by Kittyhawks as 3 Squadron RAAF advanced westwards. Bobby's plan was to have it shipped to Australia as a war trophy, but he was thwarted by a signal from Headquarters Middle East ordering him to deliver it for performance trials. It was the first of the new Bf 109 G aircraft captured in serviceable condition, and the type had been a real headache to the Allies, outclassing all opposition including the Spitfire V. Somewhat reluctantly, he ferried it east to Heliopolis, Cairo.

Bobby provided more useful information in the shape of a photograph of the machine taken the day it was captured. It bore a simple two colour camouflage, white North African 'theatre markings' and a black '6' on the rear fuselage. Disappointingly, there was no evidence of a Luftwaffe unit marking.

It was very little to go on, I admit — a serviceable G-2 captured in Cyrenaica. There must have been many others. The trail went cold until, 11 months later, yet another picture of CV-V appeared in *Air International*. Submitted this time by Frank Smith (and again misidentified!) the aircraft

Left *Bobby Gibbes investigates the lot of the footslogger in 1942.* **Right** *Bobby after a combat-sortie.*

Bobby Gibbes prepares for the first flight in his prize. Damage to the prop blade is visible, as is the repainted red spinner and lower cowl, probably in grey.

was shown at Lydda, Palestine in 1943. I now knew where it had gone after Bob Gibbes delivered it to Cairo.

Unbeknown to me at the time, Ross Butler of Rolls-Royce had been seeking information on the Daimler-Benz engine in the portals of the Public Record Office at Kew. He had had some success and suggested that I might continue my search there. As soon as we could arrange it, Ian and I set about feeling our way through dozens of files which may just have offered further clues. Most of the day we read snippets on captured Messerschmitts,

A Bf 109 F is fuelled on a grass strip in Russia.

most aggravatingly unidentified. Crashed enemy aircraft reports were more revealing and it was exciting to learn that many aircraft from the same production batch as 10639 had been captured in the desert in late 1942 and early 1943. We were more confident than ever that we were on the right track. Late in the afternoon, I flipped open the battered card cover of the umpteenth file, and I could not believe what I saw! Gummed on the inside cover was a signal, headed 'M.E. EAI/580 of 10.1.43. Me 109 G.' which began: 'The chief differences between the Me.109F-4 and the Me.109G-2, No.*10639* being tested at Lydda, are noted below: ...'! It was hard to contain our excitement in the hushed atmosphere of the reading room!

It was to be some weeks before Ian and I could return, but we began by exploring other avenues, the first of which was obvious. We carefully read the pages of the Lydda Ops Record Book, and there we discovered the link. 10639 had been flown to Lydda from Heliopolis by a Group Captain G.M. Buxton to begin performance trials. Subsequent entries gave details of test flying and the information so urgently required by the Allies (see Appendix B). The trail again grew cold when the Ops Record Book declared that Buxton had flown the aircraft to Shandur on the shore of the Great Bitter Lake at the southern end of the Suez Canal system on 21 February 1943. Ian and I searched for hours in all the files relating to this airfield and all the units stationed nearby, but we found no reference to our 109.

As may be imagined, it was extremely frustrating to find our line of research curtailed. But at least we had proved, beyond all shadow of doubt, that 10639 had never seen Sicily, let alone been captured there! We did not have long to wait for our next lead, and it was the result of a quite remarkable coincidence. One afternoon, in our dismal shed at Northolt, we showed Chris Wills over our 'baby'. In conversation, Ian and I related our recent progress in tracing its history and we mentioned the name Buxton. 'I know a Group Captain Buxton,' he said quietly! I can scarcely describe our amazement. Chris's father, Philip, had been a leading sailplane pilot. A schoolfriend of his with similar interests had designed a sailplane and the pair commissioned its build in 1934. That friend was Mungo Buxton, then a Squadron Leader in the RAF. Chris provided me with an address in Norfolk

In centre, Group Captain Mungo Buxton. Note that an 'F'-series canopy is fitted.

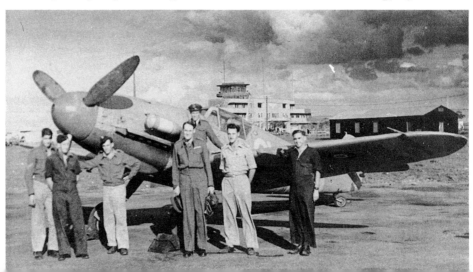

Heinz Lüdemann in happy mood in Russia.

and I was soon corresponding with the very officer who had tested our aeroplane in Palestine. His logbook entries broadly confirmed the information we had found at Kew, but of the flight to Shandur there was no mention. In fact, some weeks before he was reported to have made the delivery, Buxton departed Lydda in a Hurricane to take up his new appointment as Officer Commanding Khartoum.

We turned our attentions at the Public Record Office to the career of the aircraft after its arrival in England. The Ops Record Book of No.1426 Enemy Aircraft Flight was very revealing, and provided day-by-day detail. Through the good offices of Ken West, author of *The Captive Luftwaffe* (sadly out of print), entries from the logbooks of Flight Lieutenant E.R. (Lew) Lewendon, the Officer Commanding, and Flying Officer Doug Gough, embellished the picture. The British career of RN228, as it was then known, was comparatively easy to compile.

Still we had found nothing of the Luftwaffe career of the *Gustav*. I bemoaned the lack of a unit marking on the aircraft which would have pro-

Heinz Lüdemann contemplates his next move .

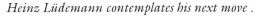

vided a useful lead. Some years were to elapse during which the odd photograph of 'CV-V' emerged. My attention necessarily turned to the reconstruction of the aeroplane and I could afford little time to continue the search for historical detail. At this point Andy Stewart stepped in and carried the investigation further. He contacted the *Bundesarchiv* in Koblenz, requesting any information that surviving records might reveal on the fate of *Werk Nummer* 10639. The answer was just what we had been looking for. Bf 109 G-2, *Werk Nummer* 10639 was lost through 'enemy action' at Gambut airfield on 11 November 1942. Its pilot, Leutnant Heinz Lüdemann, received slight injuries in combat. A further letter confirmed that 10639 had, indeed, been *'Schwarze 6'*. Heinz was a pilot with the 8th Squadron of the 77th Fighter Group — in Luftwaffe terms, 8 *Staffel/Jagdgeschwader* 77, or 8/JG77.

The next step was to try to locate him, and through a friend, Rick Chapman, then assistant editor of *Flugzeug* magazine in Germany, we made contact with Heinz Langer, Lüdemann's nephew. Lüdemann, unfortunately, was killed in combat with P-40s near Ksar Rhilane in Tunisia on 10 March 1943. He was 22 years old. Until mid-October 1942 he had fought in Russia. His unit was withdrawn from the eastern front for redeployment to the desert, and they flew their Bf 109 G fighters to Munich. Following some home leave, they reported to München-Riem airfield to pick up new *Gustavs*. The first stage of their journey took them to Bari in southern Italy, a large distribution base for the Luftwaffe. Over the course of a few days, the radio codes (*Kennzeichen*) on each aicaft were washed off and replaced by their new *Staffel* numbers.

Lüdemann finally arrived in North Africa on the afternoon of 31 October. While neither his logbook nor his diary reveal which aircraft he flew from Germany to Africa, it would appear that his normal steed had been 'Black 4'. However, it is a reasonable assumption that 'Black 6' flew the same route

Left *A quiet read between sorties in Russia for Günter Behling (left) and Heinz Lüdemann.* **Right** *Heinz Lüdemann in dress uniform.*

Top: *10639 at Lydda, Palestine minus its canopy which came adrift on take-off. Note the fuselage band has been over-painted.*

Above: *CV-V leaps into the African sky, the white fuselage band still undisguised.*

Below: *Lüdemann and his 'sohn' — his 'son'!*

Top: *451 Squadron detachment at Lydda. On aircraft: Sgt Harold Osborn and Sgt Bill Fisher. On ground (left to right): Hec Christian, Don Batger, Brian Arnold, Sam Jordan, Joe Branch and Lloyd Barry.*

Above: *Gordon Jones on the boss's 109. This is the shot which revealed the true markings on the wing's leading edge.*

Below: *Another view taken shortly after arrival at Lydda. Behind is a Junkers Ju 87 D which was used by 601 Squadron to ferry (liquid) supplies!*

and possibly on the same days. His normal aeroplane unserviceable, he flew 'Black 6' on 4 November from Quotafiya, just west of El Alamein, and was slightly wounded in combat with P-40s. The previous day the German retreat had been ordered, and Lüdemann left 'Black 6' behind, travelling west by road in the chaotic conditions caused by the Allied breakthrough. But, as we know, the *Bundesarchiv* records declare its loss some seven days later at Gambut, some 200 miles away. Being only slightly damaged, it had been ferried there to be repaired, but such was the speed of the Allied advance, it was abandoned following removal of vital equipment.

Some years later into the research, Andy Stewart had dug long and deep into the Allied records of air activity following the start of the Battle of El Alamein. Using an extract from Heinz Lüdemann's diary, he thoroughly examined all Allied combat records for 4 November. Heinz wrote: 'During an attack on an English bomber force today I received scratch wounds to my head and body from the fighter escorts. However, I managed to get my machine home.'

Only one report in Allied records ties in with the day, time, form of combat and damage to the Bf 109. Filed by 211 Group, it revealed the following: 'Eight P-40s of 64 Squadron (USAAF) flying close cover and eight P-40s of 64 Squadron escorted 18 Baltimores which bombed enemy concentrations from 8,000 feet at location 852 297. Two red flares were seen before four Me109s and two MC202s attacked close cover, and top cover dived into attack.' In the combat which followed, Lieutenant Roy Whittaker destroyed one of the Macchi fighters and claimed a 109 damaged. The only other person to claim damage to a 109 was Lieutenant Weaver from the same Squadron. No other combat of the day comes remotely close to the description given by Lüdemann. We feel, therefore, that either Whittaker or Weaver caused the damage to 10639 that was subsequently discovered at Gambut. Unfortunately, Whittaker passed away in 1989, while Andy has yet to trace Lieutenant Weaver.

The movements of the aircraft over one short period remained obscure. If it did not fly to Shandur, where *did* it go? The answer was finally supplied by

A clean side view showing its new canopy and markings.

10639 in front of an American B-24 Liberator.

*After arrival at Kasfereet, the 109 has a new propeller and spinner and yet another 'F'
canopy. The rear glazing has been crudely repaired.*

10639 poses in front of Spitfire Vc EP982 prior to tactical trials.

Following painting at Collyweston in February 1944, 'Lew' Lewendon stands by the nose while Doug Gough leans on the wing.

Lovely view of another Bf 109 G-2, W.Nr. 10533 of I/JG77. The fate of this aircraft is unknown.

Wing Commander Dick Martin. As Unit Test Pilot of No.107 Maintenance Unit, Kasfereet, he had flown a certain Bf 109 G, coded CV-V. Kasfereet was a huge airfield on the shore of the Great Bitter Lake, not far from Shandur and it was the centre for repair of mainly American equipment, and also the reception base for such machines arriving from the factories. Dick flew 10639 only once, but there was at least one other flight as he pitted a Spitfire V against it in tactical trials. From what he remembered of Shandur, there is little likelihood that the 109 had been flown there. I surmise, therefore, that the compiler of the Lydda Ops Records had been ill-informed and that the aircraft flew to Kasfereet. Following completion of the trials, it was dismantled, crated and shipped to England.

I have no doubt that more detail will emerge on the life story of 'Black 6'. Nevertheless, a fairly full account of its career has been uncovered and more detailed information is provided in Appendix A.

17

To Duxford

'**B**lack 6' had become the responsibility of the Imperial War Museum in an agreement due to last a scant three years. It was important to Duxford, particularly as the display season had already begun, that it was moved there with all due speed. With this very much in mind, my team immediately prepared the engine for removal.

We hoped to save the vast quantity of water glycol mixture by draining it in controlled fashion into special drums. Ian and I had some difficulty easing a large rubber hose from a flange. The idea was to partly remove it and allow the coolant to trickle into the container beneath. After its initial resistance,

Partly stripped engine. The right cylinder bank has been removed showing the large pistons.

Above left: *Sump cover removed, the crankshaft and its balance weights lie available for inspection.*

Left: *Paul Blackah works on the bottom of the engine. The propeller reduction gear has been removed.*

Above: *Inside view of gear wheels of propeller reduction gear.*

Below: *The basic engine block in its roll-over frame.*

the hose slipped off with remarkable ease and I directed the torrent towards the container. I think about one half reached the intended target, the remainder saturating my trousers and shoes before ending up on the hangar floor. As for Ian, he was some yards away, chortling with all the rest! I have never seen him move so quickly!

One cylinder bank lies on the bench.

Arrangements were made to transport the roll-over frame from Bristol which had been used in the engine rebuild. It arrived a few days later and was promptly shot-bolted to the floor. The Daimler-Benz was then removed from the airframe, and a stripdown began under the guidance of Roger Slade.

In addition to the exhaust valves presented by Oskar Friedrich, I received

The starter mechanism complete with articulated crank shaft. The small flywheel is on the right.

One engine exhaust valve showing the damage caused by loss of its chromium plate.

Black 6 crosses the perimeter of its new home, Duxford, flown by John Allison, 12 July 1991.

another from Günter Leonhardt, and a pair via a friend in Paris, Jean-Michel Goyat. Soon afterwards, however, I located a full replacement set. I was aware of the extent of the damage to one valve, but it was likely that others had also suffered. I had to know the reason. Once removed, we found that the chromium plate applied to the valves had lifted on several. With the pro-

The team prepares Black 6 for a display at Duxford.

The starter being wound by Paul. On the left, Ian Mason gets ready to remove chocks.

tection missing, the extreme heat of combustion had slowly burned away the metal. Chris contacted the Stellite Company of Swindon which concluded that the method used in removing the old plate had been incorrect. (It had

John Allison taxies out for a test flight.

The author about to enjoy himself!

been ground off mechanically, rather than destroyed chemically.) The new plate, therefore, had little hope of adhering to the abused valve surfaces. Stellite willingly undertook to examine our replacement valves and we heard

Black 6 in front of a Hispano HA1112 Buchon.

Our baby poses prettily in front of the control tower at Duxford.

a few days later that they were all in perfect health.

I was to play no part whatsoever in the engine repair work as I had moved home from Northamptonshire to Ayrshire in Scotland. I received regular reports from Paul who made time virtually every day to advance the work, usually aided by Bob Kitchener. The others were in regular weekend attendance. It was as close to an all-out effort that our volunteer status would allow, and I was immensely proud of them all.

A view not far removed from that taken on its capture in 1942.

Left *John Dixon makes an adjustment watched by Paul Blackah while Ian Mason busies himself on the wing.* **Right** *A beautiful day at Duxford. Ian Mason helps Paul Blackah wind the starter.*

The engine was refitted by the end of June. Following reconnection of all systems, and replenishment of oil, coolant and hydraulic systems, we were ready to test it. On the first weekend in July Roger performed the first run. Very reluctant to start (at least that had not changed!), it behaved very well,

A lovely view of the magnificent DB 605. A warning about wheel size is nicely highlighted on the leading edge of the wing.

Black 6 rests between test flights following a heavy shower.

and even sounded like its old self.

John Allison decided that test-flying would not be carried out at Benson because of the appalling condition of the grass strip. Various suitable airfields were considered, but as Duxford had one of the best grass runways — and

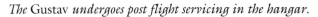

The Gustav *undergoes post flight servicing in the hangar.*

Left *The VDM propeller at rest.* **Right** *The author chats to Doug Gough and Ian Mason on a misty autumn morning.*

the 109 was overdue there anyway, he decided it should be flown there directly to continue the test programme. The Classic Fighter Meet, the big show of the year, was seven days away. I informed IWM that I felt confident

The exhaust stain blankets the starboard wing-root.

Three bodies anchor the tail while Roger checks the engine in the English mist.

the aircraft would be ready for a ferry flight within days and, happily, we discovered that John was available on the Friday. Coincidentally, I was at Luton for my six-monthly simulator ride on the Monday and Tuesday, but free the following two days. Ian picked me up from my hotel and we drove to Benson where the 109 was readied for John's first flight. First of all, a full check of the engine and systems was necessary. I was not prepared to clear

Nose view of the Gustav *during an engine test.*

Chris Starr contemplates the Green Shield stamps while Graeme pours in more fuel.

Major Fritz Losigkeit, formerly of JG77, gives his approval of our restoration work.

the aircraft for a cross-country flight, even a short one, unless everything was 'tickety-boo'.

I had to satisfy myself that the rebuilt engine was in good fettle, naturally. My first test revealed a moderate mag-drop which I attempted to burn off with more power than normal. It did not work, and once again we resorted

A Bf 109 G taxies past a P-51D Mustang, its principal opponent later in the War.

Paul Blackah begins a pre-flight check of the engine. The supercharger impeller is hidden in its circular intake.

The Luftwaffe arrives! Left to right: Helmut Rix (JG301), Fritz Losigkeit (JG77), Kurt Wuppermann (JG54), Christian Dezius (JG2), Julius Meinberg (JG2) and Herbert Thomas (NJG).

The 109 looks like a model as it lands watched by some glider pilots.

to changing a few spark plugs and cleaning others. On Thursday the old bird
appeared in very good condition and I took the opportunity to introduce

The distinctive lines of the Messerschmitt shown off by Dave Southwood.

A busy scene at Duxford. The tail is tied to the concrete prior to an engine check.

An interesting shot of the engine minus its oil tank, and revealing the complex pipe-work.

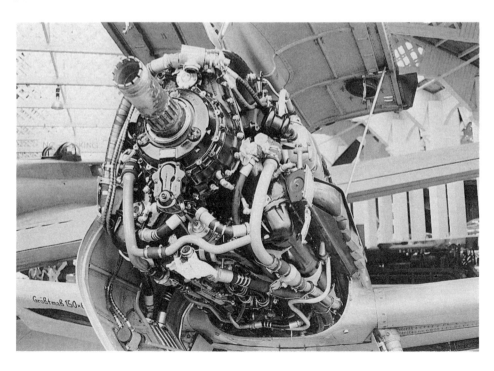

The end of another successful day. Black 6 is towed back to the hangar.

The propeller is refitted by Paul Blackah and Ian Mason. Howard Cook is partly hidden behind.

Bob Kitchener inspects the oil thermostat for leaks. The exhaust stubs have been fitted with protective plugs.

Beautiful view of Black 6 over Cambridgeshire. The white spinner and wing-tips always stand out well.

both Ian and Chris to engine-running a DB 605. I needed other members of the team to be qualified in engine handling as I was unlikely to be readily available throughout the display season.

Regrettably, my employers had need of me the following day and I was unable to be present to watch 'Black 6' depart from Benson. On the evening of Friday 12 July, though, I learned from several phone calls from the team that it had landed safely at Duxford. A largely uneventful flight, John Allison was apparently pleased with the feel of the aeroplane. It has never returned to Benson, landing on the grass strip being considered too risky for such a valuable machine.

Although it could not fly at the Classic Fighter Meet, the 109 was the focus of attention of the large crowd. Parked close to the barrier, the Team took the opportunity to show off the beautiful engine by 'opening the hood'. Twice during the day, the second time by popular request, it was started and the mighty Daimler-Benz made its distinctive deep growl known to thousands of people. A German friend of mine described it perfectly; it is a sound which fills the sky. Our modest debut was very well received.

John Allison was not available to begin the test programme stipulated by the Civil Aviation Authority until the beginning of August, but at least the delay allowed most team members to arrange their respective affairs in order

Black 6 flown by Dave Southwood is escorted by a restored Spitfire 16 in the hands of John Allison.

A moody view of the 109 with its radiator flaps wide open.

to be available. Testing was completed in six flights; an incredibly short programme for an aircraft which had been grounded for 46 years, and a tribute
to the skill of my colleagues. Only one adjustment was required; the control
surface tabs needed slight trimming. And we had one failure — the drive to
the rpm gauge.

*Duxford 1992, and 1426 Flight joins the author for the camera. Left to right: Bernard
Albon, the author, 'Lofty' Westwood, Doug Gough and Raymond Frisby.*

John Allison gives the team a close look at their baby in flight.

It was a very happy couple of days for us all, and even frequent heavy showers did not dampen our high spirits. It was not unusual to see the entire team sheltering under the 109, shuffling from one spot to another to avoid the intrusion of yet another stream of rain. After one such downpour, we were surprised to find that the narrow gap between the windshield and the armour glass had partly filled with rainwater! John flew the aircraft, water swilling in front of his eyes, to explore the effect of controls. During the flight, one manoeuvre induced sudden negative 'G'. Describing the event to us afterwards, we could not understand why the water had not been ejected!

Although it did not affect testing unduly, the nagging occasional drop in rpm when checking the magnetos remained. To our relief the fault was traced in the start switch by an electrician at Duxford. There were to be no further worries with the ignition system of the Daimler-Benz.

The test details were quickly submitted to the CAA, together with our proposed Pilot's Notes, and we received our Permit to Fly a few weeks later. The first flying display of 'Black 6' occurred in mid-September when it was beautifully flown by John through a sequence of restrained aerobatics. I could not have wished for better.

18

In conclusion

W as the exercise worthwhile? It was an awful lot of hard work, but I believe that it was. Both project aims have been achieved. Primarily, 10639 is now the most extensively, and correctly, restored Messerschmitt Bf 109 to be found anywhere — a far cry from the derelict, gutted and abused hulk of 1972. My work has ensured that all markings found internally have been saved and repaired where possible. A host of others have been carefully reproduced through close scrutiny of the few photographs which miraculously survived. It remains a deep disappointment to me that the external paintwork was so unthinkingly destroyed at Wattisham. Nonetheless, despite a total lack of such valuable evidence, the 109 has been returned to its correct appearance using paint mixed with reference to original paint chips, and the markings, dimensionally correct according to the relevant Luftwaffe manual, have been accurately located, again with reference both to the appropriate documents and photographs.

The secondary purpose of the project has also been achieved, confounding the opinion of many. The return of the Bf 109 to airworthiness was largely responsible for the extraordinary length of the restoration, but it was by no means the only reason. I will concede that much more work was performed on the airframe than was strictly necessary in an engineering sense. However, had I not completely stripped the aeroplane, for instance, the restoration would have been badly compromised. As a result of such in-depth examination, every inch has received attention. Moreover, and most important, it remains the *original* machine. All too often, recent 'restorations' have culminated in new-build reproductions which, though marvellous pieces of engineering, can scarcely be claimed to be the original aircraft restored.

Would I do it again? No — certainly not on the same basis. (I would be well into pensionable age, for one thing!) The Bf 109 project suffered unnecessarily from an almost total lack of official support. There is no denying that I volunteered to restore it, of course. (It was the one occasion in my Service life where I ignored the sage rule — NEVER volunteer!) But the

The supercharger intake almost hidden, the clean lines of the little Gustav *are apparent, as is the value of the bright yellow cowling as an aid to identification from the ground.*

absence of funding, although tolerable, made life extremely difficult, and many valuable and rare parts on offer over the years had to be rejected. Lack of authority to approach workshops for help, for example, rendered my task exasperating. I could not reconcile the obvious anomaly of an MoD-'owned' aeroplane being restored on MoD property where facilities were denied more often than not.

The dreaded white nose. A white propeller spinner meant only one thing to Allied pilots in the Desert — the enemy, and more often than not, a Messerschmitt Bf 109.

The Messerschmitt was restored on a shoestring. It must not be allowed to happen again. The trouble is that, lest there are changes made, I can see little hope of the situation improving. I have had some years to contemplate the lot of the dozen other German aircraft which remain in the charge of the Ministry of Defence. Whilst I do not pretend to harbour a plan which will overcome the many hurdles standing in the way, it would be remiss of me not to urge an awakening from the lethargic attitude of the past, and the present, and suggest that a programme of restoration should be contemplated.

My particular interest lies with the German aircraft in the United Kingdom, of course, and I recognize that they represent but a small segment of the overall picture. However, they have suffered most from neglect. It is worth recalling how they came to be on display at Hendon and Cosford. Until the late sixties a large number of historic machines survived, cared for by the Royal Air Force on behalf of the Air Historical Branch of the Ministry of Defence. The creation of the Royal Air Force Museum provided a home for many. The lot of the remainder, including the Luftwaffe equipment, was unaltered, although responsibility passed to the newly-created Historic Aircraft Committee. Subsequently, five were given shelter in the Battle of Britain Museum at Hendon and, later still, two smaller fighters were shoe-horned into the Bomber Command Hall, whilst the larger machines were sent to Cosford. It is a sad fact that little maintenance has been afforded these aircraft since, and none has benefited from any restoration work other than through voluntary effort. The majority of those at Hendon bear paint which was hurriedly applied to make them presentable for display over 20 years ago. Internally, most are in abysmal condition.

Although on display at Hendon and Cosford, all German machines are on loan from the Ministry of Defence. It may be naïve of me, but I have often wondered why a government body, charged with the defence of the Realm, should continue to be responsible for 50-year-old historical relics! The inadequacy of the arrangement is illustrated by the fact that the military members of the HAC fulfil their roles as 'secondary duties', i.e. on a part-time basis. Although they are able to draw on the advice of the RAF Museum, none has ever had any experience of, or interest in, the preservation of old aircraft. Furthermore, the Chairman, normally of Group Captain rank, is frequently changed on posting, and continuity is dependent on a more junior officer. Inevitably, his successor requires time to 'acclimatize', but, with all due respect to the officers in question, he is a novice with a temporary role to play.

I am far from convinced that the RAF Museum, on the other hand is the ideal keeper. Undoubtedly hampered by insufficient funds over the years, nevertheless we have seen the purchase of expensive replicas, whilst the small staff at Cardington battle to restore original machines. The latest disturbing trend has been the disposal of some exhibits and scrapping of others, one as a result of utter neglect. While I have misgivings about placing the German collection in the hands of the Museum, there can be no argument that these historic aircraft should be the Museum's responsibility. However, it would be essential, to my mind, that irrevocable safeguards are first agreed. The most important being that, under no circumstances, will disposal ever be contemplated.

What could be achieved, given the will and the money. The Smithsonian Institution's Focke-Wulf Fw 190 F-8/R1, W.Nr. 931884.

For a number of years, the Smithsonian Institution in the USA has been following a disciplined programme of restoration. Each aircraft in turn is researched before a painstaking stripdown is initiated. All markings, however small, are recorded, and external paintwork is carefully rubbed down to uncover original markings. These are drawn, photographed and colour matches obtained. The aircraft is then totally stripped, cleaned and repaired, whence the rebuild begins. Each and every item is accurately restored and conserved and a full complement of equipment installed, after which the subject is restored to its original appearance. (Note — *original* appearance, not the prettiest colour scheme the restorer favours.) The results in Washington have been truly magnificent.

I strongly advocate that the RAF Museum emulates the Smithsonian, certainly where the Luftwaffe aircraft are concerned. But I cannot stress too strongly that patient, in-depth research is first required before such work is contemplated. It is also essential that the people tackling the work should be knowledgeable and motivated. Contracting such work would result in a second-rate result; it will only be as good as the money offered.

Finance is undoubtedly a problem, though, and always has been. The time has come, however, for money to be spent on a restoration programme rather than on enlarging the national collection. Treasury funding should be devoted specifically to a restoration task, rather than allocated as a result of a vote by committee. To date, the official approach has been one of minimum possible maintenance for these valuable and rare machines. It is simply not good enough. They deserve, and need, better from their keepers.

As for 'Black 6', the restoration continues. I still search for many more parts with which to enhance the aeroplane. Unfortunately, little time remains. At the end of October 1994, the unique agreement between the Ministry of Defence and the Imperial War Museum is due to terminate, and

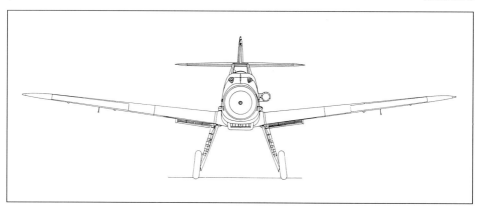

the Bf 109 G will 'revert to departmental control', to use military jargon —
the length of its flying career, and its fate, having been decided arbitrarily in
conference in early 1990. It would be foolhardy to suggest that 'Black 6'
should be kept in the air indefinitely. It is, after all, a man-made machine
and, like any other, has a finite life. Nonetheless, whilst chronologically 50
years old, the airframe is young in terms of usage. Most important, it is lov-
ingly pampered and cared for by a dedicated team.

It will have flown for three summers amassing less than one hundred
hours in the air after a rebuild which spanned 20 years, and the knowledgeable enthusiasts responsible for its resurrection will disperse. I am bound to
say that I am saddened by the prospect.

Appendix A

A Chronology of 'Black 6'/*W.Nr.* 10639

Sep 42	Built at Leipzig by Erla Maschinenwerk GmbH, probably at its plant at Mockau. Construction started as a Bf 109 F-3, but aircraft converted to Bf 109 G-2 standard. Radio code allocated PG+QJ.
13 Oct 42	Accepted by the Luftwaffe.
21 Oct 42	Collected by III/JG77 at München-Riem airfield. Flown to Vicenza, north Italy, and then to Jesi.
22 Oct 42	To Foggia and thence to Bari. While there, the radio codes were removed and the identity 'Black 6' applied.
27 Oct 42	Bari to Elefsis, Athens. In the evening, ferried to Tobruk East airfield, Cyrenaica.
28 Oct 42	Tobruk to El Harun.

Bir Dufan, December 1942. To the right is Heinz Lüdemann. In the centre a captured 'English' sergeant pilot with slight facial burns. Who is he?

The engine of 10639 in 1942, machine guns still in position.

2 Nov 42   El Harun to Bir el Abd. It is probable that aircraft was flown from this date by Leutnant Heinz Lüdemann; his aircraft, 'Black 4', having been damaged the previous day.

3 Nov 42   Bir el Abd to Quasaba on the date when the German retreat was ordered. Quasaba back to Bir el Abd.

4 Nov 42   Bir el Abd to Quotifaiya. Take-off for combat. Entry in Lüdemann's diary. 'During an attack on an English bomber force, I was slightly injured in my head and body by the fighter escort. However, I managed to get my machine home.' Aircraft ferried, probably immediately, by an unknown pilot to Gambut Main airfield, south-east of Tobruk, for repair.

13 Nov 42   Discovered by Flight Lieutenant Ken McRae, Engineer Officer, 3 Squadron, RAAF at Gambut Main. Had been 'shot up'. Damage to tailwheel, tailplane, canopy and one prop blade. Radio and oxygen unserviceable and some instruments missing. Sqn Ldr Gibbes estimates that it had a low number of hours, 'perhaps no more than ten, and these had probably been logged during its ferry to the Western Desert. It was not operational and had not been fitted with reflector sight or armament.' (In fact these, and other parts, had been removed.) 3 Squadron History states: '10 Nov 42 — Whilst at LG101 the CO located a Me109G, the latest type of fighter in service with

Gambut airfield, 1942. CV-V is prepared in front of a captured Bf 109 F.

	the Luftwaffe.' Both date and location are erroneous.
14 Nov 42	New tailplane, tailwheel and canopy fitted, and hole in prop blade plugged. Worked until after dark putting on 3 Sqn marks. (CV-V, the personal code of Bobby Gibbes.)
15 Nov 42	Gibbes flew CV-V to Gazala Satellite II escorted by Kittyhawks ET899 and AK626. A long taxi in crosswind, following a guiding jeep, caused brake fires.
19 Nov 42	Gibbes flew CV-V to Martuba, escorted this time by Kittyhawks ET899, AK626, ET951 and FL323. From his diary: 'The 109 is a hell of a nice kite with a terrific performance. On lowest permissible boost and revs was clocking 220-230 mph.'
21 Nov 42	Gibbes diary: 'In the afternoon took up 109G for an RAF film unit. Only up 10 minutes as battery a bit flat and prop motor wouldn't work.'
1 Dec 42	Two familiarization flights, one by Gibbes and the other by Flt Lt R.J. Watt. Gibbes tried cannon out. Afterwards cannon and reflector sight removed. Excerpt from '3 Squadron at War': '... a somewhat mysteri-

CV-V being serviced by a detachment of 451 Sqn.

ous signal was received from HQ Middle East stating that the captured aircraft had to be flown back there.'

2 Dec 42   On instructions of AOC, Gibbes flew to RAF Heliopolis. Route: Martuba, El Adem, Mersa Matruh, Amiyra, Cairo. 'On ferry flight back to Delta, I carried out a couple of flypasts of Dakotas, again with interesting results when the pilots saw a 109 passing them. It was mean but at the time amusing, if only to me.' On take-off from Amiyra, the canopy flew off, hitting starboard wing.

4 Dec 42   Engineering detachment from 451 Sqn RAAF based at St Jean, Palestine, started work on aircraft.

5 Dec 42   Work continued. (From diary of Heric Christian, Engine Fitter IIE.)

6 Dec 42   'Got 109 going.'

7 Dec 42–   109 remained at Heliopolis while 451 visited El Daba, Mersa
12 Dec 42   Matruh and Quasaba seeking spares.

13 Dec 42 'Working on 109 all day and found out a few more things about it.'

14 Dec 42 'Worked all day on plane and ran it up to full revs.'

15 Dec 42 'Up early and got plane ready and Gp Capt (Buxton) flew it to Lydda.' Buxton's log gives date as 16th, but this is contradicted by Lydda Ops Record Book which states that on the 15th the 'arrival created a great deal of interest.'

19 Dec 42 'Worked on 109 all day.' Heric Christian, 451 Sqn.

20 Dec 42 Ditto.

21 Dec 42 'Plenty to do all day on 109.'

22 Dec 42 'Nice weather, plenty to do.'

23 Dec 42 'Bags of work on ME and is the motor hard to work on — I'll say.'

24 Dec 42 'Worked on 109 till 4pm.'

27 Dec 42 'Cold and windy conditions — too bad to work.'

28 Dec 42 'Worked on 109 and ME now ready to fly.' 451 checked or changed the magneto, changed the oil and filters and plugs and the rudder. The oil radiator flap was locked open, as received, because of a malfunctioning thermostat.

29 Dec 42 First test flight by Gp Capt Buxton checking speeds and rate of climb.

30 Dec 42 Second test flight, on a cloudy day, to check speeds and climb to 25,000 feet. 'Very good performer.' From Don Batger, 451 Sqn: 'He turned the 109 inside out and came back and said that it was better than anything we had at the time.'

31 Dec 42 Third test flight involving climb to 32,000 feet. 451 Sqn then handed aircraft over to Lydda Communications Flight.

17 Jan 43 Fourth test flight for partial climbs to 5,000 and 20,000 feet. Before this flight, the propeller was changed.

19 Jan 43 Fifth test to check the effect of radiator flaps, etc., on speed. A sixth test flight was also undertaken this day but it was cut short when the canopy flew off.

28 Jan 43 An extra flight was made, it is believed by Ronald Harker of Rolls-Royce. In a letter he explained: 'While I was out in the Middle East an Me109G was captured intact. Dawson (AVM Graham Dawson) instructed me to fly up to Lydda ... and test it. I ... found it very lively but not as pleasant to fly as the Spitfire. It was very interesting to me, however, to be able to compare the two aircraft from a fighting point of view, checking manoeuvrability, pilot's view and rate of climb, turning circle and the ability of the Daimler-Benz engine to keep running when upside down.' From Buxton: 'So the performance figures I

gave Harker ... and Harker's flight in the 109 were probably the most useful products.' R-R were then working on a two-stage super-charger to boost the performance of the Merlin, this unit eventually equipping the Spitfire IX.

29 Jan 43	Seventh test to check climb performance at full throttle to 35,000 feet. Eighth test to check stalling.
Feb 43	Ferried to No.107 Maintenance Unit, Kasfereet by Wing Commander Budd for tactical trials.
21 Feb 43	Flown by Unit Test Pilot, Flight Lieutenant Richard Martin.
24 Feb 43	Flown by John Penny in simulated dogfight against Spitfire Vc, EP982 flown by Martin.
Feb 43	Also flown by WO 'Paddy' Donaldson.
26 Dec 43	Arrived at Collyweston for No.1426 Enemy Aircraft Flight from Liverpool Docks in a very badly damaged case. '... the extent of the damage to the aircraft not yet assessed.' Extract from 1426 Ops Record Book.
27 Dec 43	Aircraft unpacked, but no propeller found. (Possibly retained by Kasfereet?) Aircraft assembled using wing from a second Bf 109 G. From John 'Lofty' Westwood, Fitter IIA, 1426 Flight: '... when unpacked and laid out, ... was very untidy and damaged due to bad handling and crating in an unsuitable crate. The two 109s (another damaged aircraft arrived same day) were laid out and damaged and missing parts replaced by parts removed from the second aircraft.'
5 Feb 43	A propeller arrived from Farnborough.
8 Feb 43	Prop fitted and engine given preliminary ground run and found serviceable. Allocated British serial RN228.
15 Feb 43	Aircraft serviceable and ready for test but weather inhospitable.
16 Feb 43–	
18 Feb 43	Weather remained foul, but on 18th aircraft posed for photographs.
19 Feb 43	Initial air-test performed by Flt Lt 'Lew' Lewendon.

CV-V is prepared for packing by the Station Salvage Section at Kasfereet.

24 Feb 43	Trial flight against a Tempest V of Air Fighting Development Unit: 'but was forced to land before completing the exercise due to suspected carbon monoxide in the cockpit. It was decided to use oxygen at all times ... when the duration of the flight exceeds 45 minutes. Trials pronounced satisfactory, so there was no repeat.' A letter from Bob M. Zobell of Alberta reveals: 'My logbook indicates I flew Tempest V, JN737 on comparative turns, etc., against a Bf 109 G which took off from Collyweston. Our flying time was only 45 mins. A pilot from Collyweston flew the Bf 109 G in comparing turns, zoom and rate of roll against myself. We had almost completed the trials when he suddenly broke off and headed towards Nottingham. I followed, as a captured enemy aircraft was not to be left on its own, and suddenly his aircraft began flying in a strange manner, so I put on more throttle, finally overtaking him. We had no R/T communication so I drew up close to him and signalled for him to return with me which he finally acknowledged. Later, on the ground, a phone call came from the pilot thanking me for bringing him home. He explained that he had nearly passed out from carbon monoxide poisoning which is prevalent in 109s and his vision had become impaired so that when I flew close to him he could see me but "... didn't know where he was or what was happening." ' The pilot concerned was Lew Lewendon.
25 Feb 44	Flown by Flg Off D.G.M. Gough on picture taking sortie in company with Hudson.
26 Feb 44	Further photos taken.
27 Feb 44	Colour photos taken by *Popular Science*.
28 Feb 44	Lewendon flew trial against AFDU Mustang III. Later in same day Gough flew for Ministry of Aircraft Production photographs.
29 Feb 44	Lewendon flew a satisfactory trial against AFDU Spitfire XIV and also a further sortie for photography with a Hudson. This machine was, reportedly, too slow for the purpose and it was suggested that something fast be used subsequently. Gough's logbook refers to a trial against a Corsair on this date, but no other record of the sortie can be found.
1 Mar 44	Gough flew against an NAFDU Seafire III in the morning and a Corsair in the afternoon.
2 Mar 44	Flown for first time by Flg Off Jack Staples. Afterwards Lewendon took off for a trial against a Tempest but this 'failed to appear'.
7 Mar 44	Staples flew a trial against NAFDU Hellcat.
12 Mar 44	Unserviceable because of faulty ignition harness. Work commenced to renew harness throughout.
16 Mar 44	Air-tested by Lewendon. The engine proved satisfactory but the manual propeller control was inoperative.
21 Mar 44	Further air-test by Lewendon.
22 Mar 44	Flg Off Lewis-Watts flew for first time but bent one prop blade tip on take-off whilst trying to avoid an Airspeed Oxford. Sgt Dowie started to crop tips in order to even up blades.
23 Mar 44	Repairs completed and an air-test by Lewendon proved it serviceable. In the afternoon, a new 'tour' (No.12) began with Lewendon flying 109 accompanied by Staples in Fw 190 A-4/U8, *W.Nr.5843/PN999* and WO Gray in Bf 110 C-5, *W.Nr.2177/AX772*. They were escorted into Hullavington by two Spitfires from Colerne. A large crowd greeted the arrival and ground demonstrations followed. (Note: there

	is no record of this flight in Lewendon's log. It is probable therefore that the 109 was flown by another pilot that day.)
24 Mar 44	Took off for Bovingdon, but returned because of bad weather. No pilot's name is recorded for this flight, but it is likely to have been Lewendon — see previous entry.
25 Mar 44	Lewendon again took off for Bovingdon with the Fw 190 plus an Oxford and two Spitfires from Colerne.
27 Mar 44	Lewendon flew RN228 on photographic sortie with a Boston and afterwards gave a flying demonstration. Later, Lewis-Watts swung on take-off, damaging starboard wing-tip. The following day the Oxford was despatched to Collyweston to collect another wing-tip.
29 Mar 44	Aircraft once again serviceable but weather again unfavourable.
30 Mar 44	The tour continued to USAAF Chipping Ongar escorted by a P-38 and a P-47, Lewendon flying the Bf 109, Gough the Fw 190 and Gray a Ju 88. Air-tested by Lewendon beforehand.
31 Mar 44	Gough flew a demonstration before transitting to Stansted Mountfitchet and giving further demonstration.
1 Apr 44	Gough flew from Stansted to Great Dunmow and gave a further flying demonstration. 2 & 3
Apr 44	Held up because of weather. The condenser of one of the magnetos fell off during a ground run and was repaired and replaced pending arrival of a fresh one from base.
4 Apr 44	Flown Great Dunmow to Great Saling by Lewendon.
5 Apr 44	Awaiting new condenser.
6 Apr 44	A new condenser arrived and was fitted, but aircraft u/s because of a mag drop.
7 Apr 44	Once again serviceable and left for Earl's Colne escorted by P-51s. During a fly-over of Rivenhall trouble developed and 109 landed. Gough's log briefly records: 'Andrews Field, force-landed Rivenhall.'
8 Apr 44	U/s with undiagnosed fault.
15 Apr 44	Sgt Bill Dowie reported a cracked crankshaft web.
16 Jun 44	Replacement engine fitted. Had come from Middle East and was itself damaged by bullets, requiring a lot of attention including rewiring the ignition harness.
22 Jun 44	Air-tested by Flt Lt Dick Forbes.
30 Jun 44	Trouble found with constant speed governor on ground run. Fault cured.
2 Jul 44	Serviceable and flown by Staples.
5 Jul 44	Photographed from Boston with Lewendon flying 109. A further photo session was performed on the same day.
9 Aug 44	Left for West Raynham with Fw 190 and Ju 88 escorted by three Mosquitos.
10 Aug 44	West Raynham to Little Snoring after mock combat with three Mosquitos! Burst tyre while taxiing in.
11 Aug 44	Three sorties flown by Gough. Departed Little Snoring for Massingham. Demonstration flight given there followed by a return to Collyweston.
9 Sep 44	Forbes flew to Thurleigh in company with Fw 190 and Ju 88, escorted by Hurricane and Spitfire from Digby. Demonstration given in air and on ground before returning to base.
17 Sep 44	Gough left Collyweston for Bradwell Bay via Northolt in company of Lewendon flying Fw 190 and escorted by two Spitfires. Ground and

	air demonstrations given before returning to base.
18 Sep 44	Flight to Chipping Ongar. Instruction given to USAAF Disarmament School on maintenance and temporary immobilization of German aircraft. 109 grounded with ignition trouble.
23 Sep 44	Departed Collyweston for Leavesden, Gough flying.
25 Sep 44	Demonstration flown by Forbes. Gough then flew 109 to Chipping Ongar.
16 Oct 44	Magneto trouble at Chipping Ongar.
31 Oct 44	Chipping Ongar to Collyweston by Gough.
13 Nov 44	U/s.
14 Nov 44	Serviceable!
27 Mar 45	Gough flew aircraft to Tangmere, and aircraft transferred to Enemy Aircraft Flight of Central Fighter Establishment. (1426 Flight disbanded on 21 Jan 45.)
1 Nov 45	To 47 Maintenance Unit, RAF Sealand for storage.
1946-61	Recorded as being at 15MU Wroughton, Stanmore Park and, in the late 1950s, at Fulbeck and Cranwell. On regular view on Horse Guards Parade during Battle of Britain week 1949 till 1955.
Sep 61	To Wattisham for abortive rebuild.
30 Sep 72	Flown to Lyneham in two Hercules aircraft for start of restoration.
Jul 75	To Northolt.
Jul 83	To Benson.
8 Jul 90	First engine run.
17 Mar 91	Maiden flight.
12 Jul 91	Transfer to Duxford.

RN228 in flight displaying its RAF Training colour camouflage. Note that the fuel filler hatch has vanished.

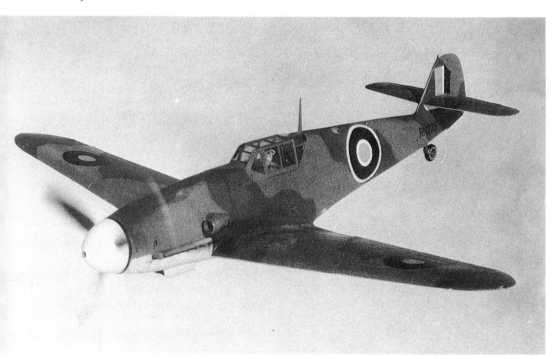

10639 in pseudo-German marks on display in Whitehall, 1950.

The Gustav *on Horse Guards Parade.*

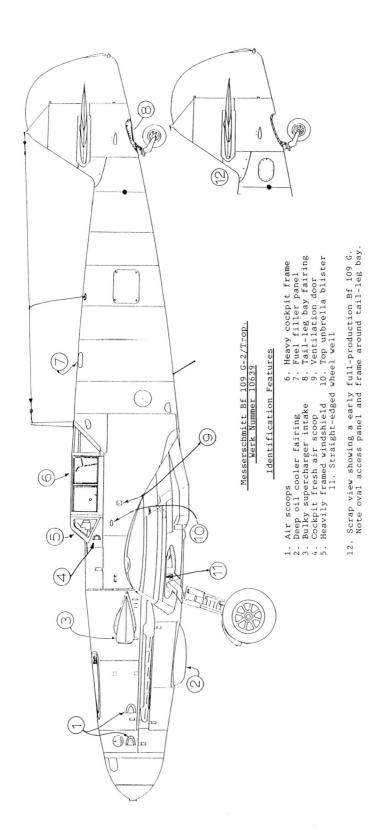

Messerschmitt Bf 109 G-2/T-op.
Werk Nummer 10639

Identification Features

1. Air scoops
2. Deep oil cooler fairing
3. Bulky supercharger intake
4. Cockpit fresh air scoop
5. Heavily framed windshield
6. Heavy cockpit frame
7. Fuel filler panel
8. Tail-leg bay fairing
9. Ventilation door
10. Top umbrella blister
11. Straight-edged wheel well

12. Scrap view showing a early full-production Bf 109 G.
    Note oval access panel and frame around tail-leg bay.

Appendix B

The cockpit of Black 6 in 1943. On the main instrument panel, the altimeter has been replaced with a British unit and the artificial horizon has given way to a Turn and Slip. Below the main panel has been affixed a British oxygen panel, whilst a large oxygen bottle can be seen attached to the floor on the right.

<u>No. 209 Group</u>.
<u>TEST OF ME.109G</u>.

C O N T E N T S.

<u>Page</u>.

<u>Figure</u>

ooooooOOOOOooooooo

TEST OF ME.109G-2 (TROP).

SUMMARY

Organisation.

1.   An Me.109G captured in the Western
Desert was allotted to No. 209 Group for performance tests.

2.   The tests were laid down in H.Q.,
M.E. letter S.54515/OPS dated November 29th 1942 and addressed to
A.H.Q., Egypt.

3.   The aeroplane was maintained by a
squadron maintenance party drawn from a Squadron in the Group, and
assistance was given by A.H.Q. Levant, particularly the Meteorological
Section, and by No.4 A.R.U.

4.   Enemy aircraft section H.Q., M.E.
gave much information and assistance, Flight Lieut. MacBean, 87246, was
attached and wrote the descriptive part of this report. He has a
remarkable knowledge of enemy aircraft.

5.   Preliminary tests were chosen to get
an outline of the performance and were completed quickly and an interim
report was sent in as required by H.Q., M.E. letter. The remaining tests
were completed by 29th January 1943.

Results

6.   The new engine (D.B.605) is little
better than the old one (D.B.601) in the 109F, the main improvement
being an increase in rated height. The fine performance is due largely
to the size of the aeroplane. It is remarkably small and light
considering the size of the engine.

7.   The cockpit is simple. A number of
technical controls such as regulation of oxygen flow, adjustment of
coolant radiator and oil radiator flaps and airscrew pitch control have
been made automatic and need no attention from the pilot. The pilot is
then able to give more attention to fighting tactics, teamwork,
navigation and practical flying.

8.   The shortcomings of the aeroplane
are, the weakness of the ailerons at high diving speeds, the weakness of
the undercarriage, the stiffness of the tail trimming gear at high
speeds, and skittishness during landing and take off.

Recommendations.

9.   The small size of the 109G remains a
prime reason for its good performance. It is recommended that British
aeroplanes should be designed to be small, but that skittishness on the
ground should be prevented by having a nosewheel undercart.

10.   British cockpits should be freed of
auxiliary technical controls which need the attention of the pilot, and
the regulation of oxygen flow, adjustment of coolant and oil radiator
flaps and airscrew pitch should be controlled by reliable automatics.

11.   It is recommended that a small enemy
aircraft test flight should be formed in the Middle East so that
performance figures are made available to Air Ministry. The personnel
required are suggested as one engineer pilot, one admin. officer, on
sergeant pilot, two fitters, two riggers, an electrician, signals and
armourer.

BRIEF DESCRIPTION OF AIRCRAFT

General.

12.   The Messerschmitt 109G-2 is a
development of the Me.109F-4, from which it is indistinguishable in the
air. The main difference lies in the engine, that in the Me.109F-4 being
D.B.601 E, while the Me.109G-2 has a D.B.605 A. There are also several
detail modifications but the airframe remains substantially the same.

Dimensions and Particulars.   Span 32' 7", Length 29' 9": Wing
    area 173 sq. ft: All up weight (as fighter without wing guns) 6820
    lbs. (approx) Wing loading 39.4 lbs. per sq. ft. (approx).

Engine. D.B. 605 A.   12 cylinder inverted 'V' liquid
    cooled in line, with direct fuel injection. This motor appears
    identical to the D.B. 601 E except for modified cylinder head
    blocks and pistons. Below are given performance figures for a
    D.B. 601 E and D.B. 605 A tested respectively in October 1941 and
    July 1942. The figures are obtained from German engine history
    sheets and the significance of the two horse power readings given is
    not known but may indicate some correction for air temperature or
    atmospheric pressure.

DB.601E. Tested October 1941.

Rating.	Revs. Per Minute	Boost. ata	lbs.	Horse Power Nx.	No.	Consumption Litres per hour.
Take off (5 mins).	2700	1.42	5.46	1425	1390	
Climb and Combat (30 mins).	2490	1.3	3.76	1236	1205	381
Max. Continuous.	2300	1.15	1.63	1017	993	298
Econ. Cruising.	1790	1.05	0.21	722	705	209
DB.605A. Tested July 1942.						
Take off (5 mins).	2810	1.42	5.46	1496	1515	
Climb and Combat (30 mins).	2612	1.3	3.76	1321	1335	400
Max. Continuous.	2312	1.15	1.63	1076	1089	310
Econ. Cruising.	2120	1.05	0.21	884	893	257

The propeller is a V.D.M.9 - 12087. Three bladed metal constant-speed with electric pitch change, hand controlled or automatic. Diam. 9' 10" Max. blade width 11 5/8".

Mechanical Features.

13.   The technical features which affect the operation of the aeroplane are similar to those for the 109F and are briefly recapitulated here for convenience. The 109 is a small aeroplane with a big engine and this largely gets its high performance. The cockpit is correspondingly small. The supercharger is driven through a hydraulic clutch in the same way as the D.B.601. This gives the effect of a multi-speed drive without attention from the pilot. The maximum boost is also automatically limited.

14.   The airscrew control can be selected for hand setting or for automatic. The hand control is a rocking switch on the throttle knob. In automatic, the airscrew governor is operated by the throttle lever to give the appropriate revs at all throttle openings, and there is no control for the pilot. The effect on range and engine life from always having the best combination of boost and rev must be good and the pilot at the same time is relieved from attending to the pitch control lever.

15.   The airscrew is electrically operated, and the handbook warns against overspeeding if a dive is started suddenly, so that the pitch change is presumably slow like our electric airscrews.

16.   The flaps of both the oil and coolant
radiators are thermostatically controlled. The operating fluid is engine
oil for the oil radiators, and hydraulic system oil for the coolant
radiators' flaps. Control of the coolant radiator flaps by the pilot is
possible but normally he will set it to automatic. The flap operation is
mechanical, by a wheel and the undercarriage retraction is hydraulic
without emergency hand pump.

Cockpit.

17.   The fuselage is clearly designed to
be as small as possible to give the maximum performance, and
consequently the cockpit is rather cramped for anyone over 6 feet tall.
The controls are laid out so that all ordinary ancillary controls are
worked by the left hand, the right side of the cockpit having only
switch buttons. This layout, combined with the automatic setting of
airscrew pitch and of coolant flaps for water and oil, simplifies the
task of the pilot.

18.   Details of the controls, which are
similar to those on some allied aircraft, are given in the German
handbook of which a translation is held in the Enemy Aircraft Section,
H.Q., M.E. A photo of the cockpit, consisting of three photos put
together, is Figure 6. The rudder pedals are level with the seat so that
the pilot is in a good position to resist acceleration; all ancillary
controls are convenient to reach and to use.

19.   Owing to the inverted engine, the top
of the front cowling is narrow and the view forward on each side is
reasonably good. The instruments supplied are:-

      Flying

         combined Artificial horizon and turn indicator

         repeater compass

         altimeter

         air speed indicator

      Technical

         Boost gauge

         R.P.M.

          Airscrew pitch indicator

          Oil and coolant temp. (combined)

          Fuel and oil pressure (combined)

          Clock.

Technical (continued).
Undercarriage lights.
Undercarriage mechanical position indicator.
Fuel warning light (20 minutes to go).

20.   The hood is small and has no curved
surfaces. The thick perspex panels are flat and allow a good view
through them. A sliding panel on top and each side allows a clear view
in bad conditions. The hood is jettisoned by a red lever on the left
side.

Armour.

21.   This is almost identical to the
Me.109F-4 and consists of one flat and one curved 10 m.m. plate
protecting the back and top of the pilot's head. Three plates, the upper
one 8 m.m. and the lower 24 m.m. protect the pilot's back. A 63 m.m.
bullet proof glass shield set at approx. 60° is mounted 13 m.m. behind
the 8 m.m. plexiglass windshield. A dural bulkhead consisting of 3Q
layers of 0.8 m.m. sheet bolted together is fitted to the lower 2/3 of
the fuselage cross section. The fuel tank is 'L' shaped and is situated
behind and below the pilot. It is of flexible rubber construction
enclosed in a plywood box for support and consists of inner and outer
layers of hard black rubber and a centre layer of soft self sealing
rubber. The capacity is 85 gallons. The oil tank is of unprotected light
alloy, ring shaped, fits round the airscrew reduction gear, and holds 8
1/2 gallons. The coolant header tanks are also of unprotected light
alloy and are mounted on either side of the crankcase. Occasionally fuel
tanks made of unprotected light alloy have been found on Me.109'Es and
F's and once on a G.

Radio.

22.   The standard German fighter R/T set,
FUG VII a, is fitted as on the Me.109F. For further details reports are
available through the Enemy Aircraft Section, H.Q., M.E.

Oxygen.

23.   The standard Draeger Unit as on
Me.109F is fitted, the supply being drawn from 3 light alloy bottles of
normal type mounted on the bulkhead behind the fuel tank. The oxygen-air
mixture is regulated by a barometric capsule up to 33,000 feet, above
which height pure oxygen is delivered, the rate of flow being controlled
all the time by the pilot's breathing, though a hand operated button is

provided to give a sudden extra supply if desired at any time.

Compass.

24.   The layout is the same as the Me.109F
and consists of a Patin Master compass fitted in the rear fuselage
behind the laminated dural bulkhead which repeats electrically to a
slave compass on the instrument panel.

Guns.

25. Gun Installation.   This is identical to the Me.109F
which has been fully reported previously, and consists of 2 x M.G. 17,
7.92 m.m. machine guns mounted over the engine firing through the
airscrew disc, and 1 M.G. 151, 20 m.m. cannon mounted on the rear of the
engine and firing through propeller hub. 200 rounds may be carried for
the cannon and 500 rounds for each machine gun. M.G. 17 are cocked and
fired by compressed air, electrically controlled. The M.G. 151 is cocked
and fired electrically. Wiring is also provided for the installation of
two more M.G. 151, 20 m.m. cannon, one under each wing, just outboard of
the wheel recesses. These guns have, however, only once been found
fitted.

26. Synchronisation.   Of the two M.G. 17's is mechanical by
flexible push rods enclosed in tubes operated by cams driven from the
accessory drive at the rear of the engine.

27. Harmonisation.   As the axes of the M.G. 17's are only
14" apart and these are only 17" above the axis of the M.G. 151 it would
appear that harmonisation is scarcely neccessary. At 50 yards there was
no apparent converging of the trajectories of the M.G. 17's but it was
not possible to check the trajectory of the M.G. 151 which had moreover
been removed and replaced by another gun making any estimate of doubtful
value.

28. Stop Butt Trials.   Since the M.G. 151 and M.G. 17 have
been fully tested and reported on by Woolwich and rates of fire and
velocities ascertained, it is not considered necessary to carry out
further trials. Reports are available through H.Q., M.E. Enemy Aircraft
Section.

29. Re-arming.   Re-arming is simple and could be
carried out by four men in under ten minutes. Operations consist of
removing empty belts and cases from cannon and M.Gs. via two quickly
detachable trays under fuselage. Lifting both sides of engine cowling
(three quick release fittings on each side). Replacing old M.G.

magazines with new and fitting belts (the M.Gs. are very accessible).
Removing cannon breach cover in cockpit (three quick release fittings)
fitting new belt (magazine in port wing accessible via two quick release
covers). Feeding belt through guide and inserting in breach. Closing
breach. Replacing cover. Lowering and securing engine cowlings. Gun
cleaning could be carried out if necessary during or in between any of
these operations.

30. Maintenance.   The guns are quickly removable for
maintenance and this should provide no difficulties. Both guns and
cannon are very simple in operation being recoil and having a minimum of
working parts.

31. Cine-camera.   Provision is made in the electrical
circuit for a cine-camera but these are not fitted as a rule and on only
two aircraft have windows in the port wing, just outboard of the wheel
recesses, been found for this purpose.

32. Sights and Sighting.   A Revi C/12/D reflector sight is
fitted (as in the 109F). This is a simple sight having no range
computing device. It is provided with a glare shield and dimming
control, and an emergency ring and bead sight attached to the right hand
side.

TECHNICAL TRIALS.

Manoeuvrability.

33.   The elevators harden up at high
speeds and retrimming is necessary, which is difficult as the trim wheel
hardens up and becomes almost sold in a dive. Some force is needed on
the stick at high speeds, but accelerations as great as the pilot can
stand can be put on

34.   The ailerons are satisfactory up to a
moderate dive, and after that were used charily owing to the warning in
the handbook of their weakness. Comparative combat trials are needed to
complete this section of the report.

Take off.

35.   The aeroplane tends to swing to the
left towards the end of the take off and firm right rudder is needed.
The take off is very good so that the throttle can be opened slowly and
it is then easier. A normal pilot in still air will take 350 yards.

Stalling Speed.

36.   Stalling speed is 102 m.p.h.
indicated flaps and wheels down, and 112 m.p.h. with flaps and wheels
up. With flaps and wheels up, the ailerons control becomes distinctly
lighter at about 140 m.p.h. when the slots open.

Trim.

37.   The trim is effective and not too
sensitive. At high diving speeds it becomes almost solid.

Landing.

38.   Approach should be made at 120 m.p.h.
indicated.

39.   The aeroplane is rather skittish and
the pilot concentrated on keeping straight and did not look at the
A.S.I. in the cockpit to see speed at touch down.

40.   From the known stalling speed, the
speed at touchdown can be deduced as 105 to 110 m.p.h.

41.   The minimum landing run to be
expected in still air with a normal pilot is 550 yards.

Performance.

42.   Climb and speed readings are
tabulated on subsequent pages. Results are also plotted as Fig. 1. Speed
Fig. 2. Rate of climb Fig. 3. Rate of climb — Different speeds Fig. 4.
Rated Height Fig. 5. Time to Height. All tests were done at maximum
engine power. (Germans have cancelled maximum emergency power,
apparently owing to engine failures.)

Dive.

43.   The dive and shallow dive are very
fast. Controls become heavy but large deflections are still possible at
350 m.p.h. indicated. N.B. The handbook warns against rough movement of
aileron controls during dive and particularly during recovery, as likely
to cause a crash. Limit speed 467 m.p.h.

Endurance.

44.   The petrol consumptions are tabulated
on subsequent page. Duration must vary greatly with operational
conditions and the time spent in dog fighting; the duration to be
expected normally is one hour.

Instrument Flying.

45.   The complete standard German set of
instruments was not fitted. Instruments flying is judged to be normal
and straightforward.

Night Flying.

46.   Cockpit and exhaust shielding are
similar to the 109 F.

Enging Starting.

47.   The engine was always started by
hand. It often started at the first attempt and always started (when in
our hands) at the second or third attempt. An experienced crew can
probably count on a start in good weather conditions at the first
attempt.

48.   A small separate tank is provided for
the priming pump which can be filled with specially volatile starting
fuel for starting in bad weather conditions.

                    (signed)   G.M. BUXTON Gp. C. (?)

ME. 109 G - 2 (TROP).

FIRST TEST FLIGHT. 0905 - 0955B HOURS 29th DECEMBER 1942

1.   The Me. 109 G. was made serviceable by 451 Squadron
Detachment.

2.   The aeroplane had been restored to standard good condition
except that the oil radiator flap was locked open as received, as the
functioning of the thermostatic control was apparently bad; no oil
thermometer was available; there was a splinter hole and score mark in
one airscrew blade.

3.   Close examination of the airscrew blade is therefore made
between flights.

4.   During take off, there is a marked swing to the left which
occurs towards the end of the run as full power is given. Acceleration
is quick and take off run fairly short.

5.   After take off, climb was started, partial climbs being made
at each height; full speed level runs were made mostly on the way down.
The figures are tabulated in Appendix "A" and the air temperatures for
compiling true speeds are tabulated in Appendix "B".

6.   The air filter was cut out at about 8000 feet and was put in
again after all tests had been done, before coming in to land.

7.   The propeller pitch control was left in automatic throughout,
and the figures of boost and revs given in Appendix "A" are therefore
those set by the automatic mechanism.

8.   The radiator flaps were left in automatic and kept the
temperature at 80°C most of the time in the easy conditions of cruising
used between test runs. The maximum temperature seen was 100°.

9.   The altimeter readings were not exact as the needle was
swinging.

10.   At 15000 to 20000 feet in the climb, the engine hesitated and
petrol pressure was seen to be low. This immediately rose on switching
on the electric pump, and the engine then ran normally.

11.   The engine has some rough periods, but usually, and
particularly at high r.p.m., runs sweetly.

12.   Owing to the high rate of climb, it is necessary to allow a
margin of up to 3000 feet to settle down into a steady climb; thus for
timing a climb from 15000 to 17000 feet it is desirable to pull up to
climbing speed (e.g. 150) at 12000 feet. In full speed level flights,
speed is picked up quickly.

13.   The control is steady and it is easy to maintain accurate
speed with reasonable care.

14.   After almost 45 minutes flying, the red fuel shortage (15
minutes to go) warning light showed. Owing to the fuel swishing about,
the light shows intermittently and is an effective warning.

15.   In the approach to landing, torque can be distinctly felt and
so the engine should be opened up slowly when "rumbling".

ME.109 G-2 (TROP).

SECOND TEST FLIGHT. 1415 - 1530B HOURS. 30th DECEMBER 1942.

1.   A series of 1000 feet timed climbs were done at various
heights. Owing to the high rate of climb, and to the fact that the
sensitive altimeter needle was swinging over a considerable arc, the
time was not likely to be accurate, and therefore climbs at certain
heights were repeated for a height difference of 2,000 feet so that any
error in timing would have less effect.

2.   The cooling flaps on the radiator are automatically controlled
and they open as the engine temperature rises. They are big enough to
cause considerable drag when wide open. Therefore they will cause more
drag on a long continuous climb and so decrease the performance. To test
this point, a long battle climb will be made.

3.   Tests are being continued in cloudy weather as the effects of
the up currents and bumps are not thought to be so important as speed in
getting some preliminary figures.

ME.109 G-2 (TROP).

THIRD TEST FLIGHT. 0920 - 1015B HOURS 31st DECEMBER 1942.

1.   On the third tests, repetitions were made of climbs and level speeds which seemed necessary to fill gaps in the performance curves plotted from the previous two test flights. A test was also made to find the rated altitude at 1.3 at: boost and corresponding revs in automatic (i.g. interconnected) airscrew control.

2.   During the climb from 30070 to 32000 feet, the cooling flaps were seen to be wide open, owing apparently to a sticky thermostat. They were adjusted to normal position by using the manual control but the rate of climb obtained must have been somewhat reduced by this.

3.   Some variation was noticed in the airscrew r.p.m. presumably due to the automatic control functioning slightly irregularly.

4.   The rated height of the engine was found by doing a steady climb at 200 mph. indicated and noting the boost pressure at each thousand feet. The automatic boost control keeps the boost constant below rated height, and it decreases above. By plotting a graph of boost against height, the line of the decreasing boost can be drawn and this meets the position of 1.3 boost at 21300 feet which is taken as the rated height as governed r.p.m. (2750).

5.   Level speed was tested at 23,200 feet (true) which gave 262 m.p.h. indicated equal to 378 m.p.h. computed.

ME.109 G - 2.

FOURTH TEST. 1505 - 1615B HOURS 17th JANUARY 1943.

1.   The original good maintenance crew under Sergeant Osborne left
with their squadron, and a new crew has taken over. No.4583 Sergeant
Dill-Franzen of No. 7 S.A.A.F. came to change the airscrew. The new
airscrew is in normal adjustment and appears to give the same results.

2.   During take off, the engine was opened up nearly to full
power, and firm right rudder was needed to stop swing to left. Take off
run is tabulated. Landing was made in reasonable tail down attitude,
nearly three point, and brakes were initially applied to stop swing, and
then both brakes were put on firmly. Landing run is tabulated.

3.   Tests were then made (at 5,000 feet) on stalling. Stall was
read first with flaps and wheels retracted, then with flaps and wheels
fully down. The figures show that 120 m.p.h. (195 km/hr) is the minimum
safe speed of approach, and only gentle turns should be made below 150
m.p.h. (240 km/hr).

4.   Partial climbs from 4,000 to 6,000 feet were then made to find
best climbing speed. Radiator flaps were set with the top flap level
with the wing flap (control on "RUHE"). An inexplicable quick climb at
160 m.p.h. was repeated and a more normal time was got. Air intake
filter was in use.

5.   Partial climbs were then made from 19,000 to 21,000 feet
(Altimeter). First 140 m.p.h. climb was probably slow as engine
hesitated owing to low fuel pressure which was raised by switching on
electrical fuel pump. A repetition was also slow as boost fell during
climb owing possibly to slight stickiness of automatic boost mechanism.
A similar slight fall of boost pressure happened at the 160 m.p.h.
climb, but not on the two others, on one of which r.p.m. was slightly
high. The intake filter was of course out for this test. The radiator
flaps were in "automatic" and were controlling satisfactorily, the
temperature being low and the opening small. The opening did not vary by
more than about an inch during this series of climbs.

6.   The engine was rather rough at times at low power, but still
is reasonably smooth at full power.

5A.   Meteorological Office, A.H.Q. Levant was asked if any currents
could occur which might account for the irregular results at 5,000 feet.
They state that an up current occurred locally between surface and 9,000
feet with a maximum of 400 feet per minute at 6,000 feet.

FOURTH TEST OF ME.109G-2 (TROP)   1 HOUR 10 MINUTES
                                         17th JANUARY 1943

Take off run   Approx. 270 yards.

Wind   12 m.p.h. 47° to take off direction.

Temperature   18.5° C. Millibars sea level 1014. Height 130
                          feet.

Landing run   Approx. 430 yards.
                          Nearly three point attitude on touching and
                          brakes used firmly.

Wind   12 m.p.h. 20° to landing direction.

Temperature   18.5° C. 1014 Millibars sea level. Height 130
                          feet.

Stall.   With flaps and wheels up. Port aileron control
                          is lost at 115 m.p.h. indicated. Nose dropped
                          at 110 m.p.h. indicated.
                          With flaps and wheels down.
                          Ailerons duff at 112 m.p.h. indicated.
                          Nose drops at 102 m.p.h. indicated.

FIFTH TEST.

Take off run.	196 yards approximately.
Wind.	15 m.p.h. dead ahead.
Temperature.	17 degrees C. 1011 Millibars sea level. Height 130 feet.
Landing run.	440 yards approx.
Wind.	15 m.p.h. dead ahead.
Temperature.	17 degrees C. 1011 Millibars sea level. Height 130 feet.

FIFTH & SIXTH TESTS.

Take off run and landing run were measured.

Partial climbs were made at 27,000 feet, which is the greatest height at which Met. Temperatures were available.

A level speed test was also made at 27,000 feet.

A level speed test was made at 16,500 feet first with radiator flaps open and then with them closed. Speeds are plotted on the speed curve to show their effect.

The sixth test, which was to be a climb to height had to be stopped before any readings were got, because the hood jettisoned itself.

ME.109. G-2 (TROP).

SEVENTH AND EIGHTH TESTS. 0935 - 1030B HOURS 29th JANUARY 1943

SEVENTH TEST.

A climb to height was made to conclude the programme of partial climbs.

The climb was made at full throttle with the airscrew in automatic.

stop watch was started quickly after taking off and operating undercarriage retracting button, i.e. about three quarters of the length of the runway. Subsequent readings were made by looking at the watch. The aerodrome is at 130 feet and the altimeter was set to 130 feet (1015 Millibars) before take off.

Between twenty and twenty-five thousand feet the coolant temperature was seen to be low and the radiator flaps were wide open. This must have temporarily reduced the rate of climb. The automatic thermostat must have stuck, and the radiator flaps were set by hand to a small opening.

To keep the (unknown) oil temperature down, the climb was made at 170 m.p.h. up to 25,000 feet and reduced to 150 m.p.h. from 25,000 to 35,000 feet.

The oil pressure was steady throughout, one division above minimum.

EIGHTH TEST.

Test of stall was made with wheels and flaps up and wheels and flaps down.

UP.   About 120 m.p.h. the aircraft began to lose height. The nose did not fall appreciably and there was no tendency to drop a wing and there was still aileron control.

DOWN.   Just below 105 m.p.h. the aircraft began to lose height but again it showed no vices and aileron control was maintained.

A slight and quick vibration was noticed when just approaching the stall. Several bursts of motor were given when near the stall but it was always possible to check any tendency to swing.

FIRST TEST OF ME.109.G-2 (TROP).   0905-0955B HOURS, 29th DECEMBER 1942.

PARTIAL CLIMBS.   ALTIMETER SET TO ZERO HEIGHT OF AERODROME 130 FT 1011 MILLIBARS. APPEARS TO BE EQUIVALENT TO 1016 MILLIBARS AT SEA LEVEL.

Indicated Altitude.	A.S.I.	Boost.	Revs per Minute.	Radiator Temperatures.	Altimeter Corrected for Calibratn.	Met. Computed Heights.	Met. Height Diffce.	A.S.I. Corrected for Calibratn.	Pitot Position Error.	Corrected I.A.S.	Time Seconds.	Rate of Climb.	True Air Speed.
6000–7000	180	1.3	2600	80 C.	5900 / 6890			181	plus 3	184	17.6		
10000–11000	180	1.3	2600	80 C.	9850 / 10850						19.		
15000–16000	175	1.3	2600	100 C.	14885 / 15890			176	plus 3	179	21.		
20000–21000	170	1.26	2700	80 C.	19955 / 20960			171	plus 3	174	19.8		
30000–31000	150	0.92	2750	80 C.	29880 / ----			152	plus 4	156	43.6		
LEVEL FLYING.													
5000	298	1.3	2650		4920	4950		298		298			322
20000	260	1.3	2700		19955	20110		260	plus 1	261	60		360
30000	223	1.3	2780		29880	29880		223		223			373 $

$ ... Using Temp. estimated by Met.

x ... Including plus 130 to allow for ground setting.

SECOND TEST OF ME.109.G-2 (TROP).

1435-1530B HOURS. 30th DECEMBER 1942.
ALTIMETER SET TO HEIGHT OF AERODROME 130 FEET 1018 MILLIBARS.

Indicated Altitude.	A.S.I.	Boost.	Revs per Minute.	Radiator Temperatures.	Altimeter Corrected for Calibratn.	Met. Computed Heights.	Met. Height Diffce.	A.S.I. Corrected for Calibratn.	Pitot Position Error.	Corrected I.A.S.	Time Seconds.	Rate of Climb.
PARTIAL CLIMBS.												
10000-12000	150	1.3			9910 / 11900	9990 / 12030	2040	152	plus 4	156	36	3430
20000-22000	150	1.3	2750		19880 / 21880	20220 / 22270	2050	152	plus 4	156	35	3510
25000-27000	150	1.1decrea-sing 1.04	2750		24880 / 26870	25330 / 27330	2000	152	plus 4	156	51.4	2360
FOLLOWING READINGS LESS ACCURATE.												
10000-11000	150	1.3	2700		9910 / 10900			152	plus 4	156	19	
15000-16000	150	1.3	2700		14890 / 15920			152	plus 4	156	18	
20000-21000	150	1.2	2650		19880 / 20860			152	plus 4	156	17	
25000-26000	150	1.1	Unknown.		24880 / 25860			152	plus 4	156	26.5	
LEVEL FLYING.												
15000	275	1.3	2650		14890	15110		275	plus 1	276		
10000	280	1.3	2680		9910	9990		280		280		
15000	267	1.3	2800		14890	15110		267	plus 1	268		
25000	243	1.14	2700		24880	25330		243	plus 1	244		

THIRD TEST OF ME.109.G-2 (TROP).   0920 – 1015B HOURS DECEMBER 1942.
ALTIMETER SET TO HEIGHT OF AERODROME 130 FEET SHOWING 1021 MILLIBARS.

PARTIAL CLIMB.

Indicated Altitude	A.S.I.	Boost	Revs per Minute	Radiator Temperatures	Altimeter Corrected for Calibratn	Met. Computed Heights	Met. Diffce	A.S.I. Corrected for Calibratn	Pitot Position Error	Corrected I.A.S.	Time Seconds	Rate of Climb
30000–32000	150				29930 / 32000	30370 / 32500	2130	152	plus 4	156	96.4	1240
22000–25000	150	1.1 / Engine Rough.	2480 & 2800 @ 25000'		21880 / 24880	22150 / 24375	2225	152	plus 4	156	77.6	
18000–20000	150	1.3	2700 increasg.		17890 / 19880	18240 / 20110	1870	152	plus 4	156	44.6	2540
4000–6000	150	1.3	2650	Rising from 90° to 100°	4000 / 5960	4020 / 6060	2040	152	plus 4	156	35.8	3420
RATED ALTITUDE.												
19000	200	1.3	2700		18890			200	plus 2	202		
20000	200	1.3	2700		19880			200	plus 2	202		
21000	200	1.3	2750		20860			200	plus 2	202		
22000	200	1.27	2750		21880			200	plus 2	202		
23000	200	1.23	2750		22880			200	plus 2	202		
24000	200	1.19	2750		23850			200	plus 2	202		
24000	200	1.15	2730		23850			200	plus 2	202		
25000	200	1.09	2700		24880			200	plus 2	202		
26000	200	1.1	2730		25860			200	plus 2	202		
LEVEL SPEED.												
23000	262	1.29	2780		22880			200	plus 2	202		

FOURTH TEST OF ME. 109 G-2 (TROP).

PARTIAL CLIMBS.

1505-1615B HOURS. 17th JANUARY 1943.

ALTIMETER SET TO HEIGHT OF AERODROME 130 FEET SHOWING 1014 MILLIBARS.

Indicated Altitude.	A.S.I.	Boost.	R.P.M.	x Radiator Temperatures.	Altimeter Corrected for Calibratn.	$ Met. Computed Heights.	$ Met. Diffce.	A.S.I. Corrected for Calibratn.	Pitot Position Error.	Ø Corrected I.A.S.	Time Seconds.	Rate of Climb.
4000–6000	140	1.3	2700	80° C.	4000 5960	4107 6101	1994	142	plus 5	147	34.3	3490
4000–6000	160	1.3	2700	80° C.	4000 5960	4107 6101	1994	162	plus 4	166	31.8	3760
4000–6000	180	1.3	2700	82° C.	4000 5960	4107 6101	1994	181	plus 3	184	35.8	3340
4000–6000	200	1.3	2700–2750	82° C.	4000 5960	4107 6101	1994	200	plus 2	202	35.8	3340
4000–6000	160 rptd.	1.3	2700	82° C.	4000 5960	4107 6101	1994	162	plus 4	166	34.	3520

AIR FILTER IN THROUGHOUT.

x .... Radiator flaps set to be level with ordinary flap.

Ø .... Pitot position error taken from R.A.E. Report No. E.A. 39/11 Brief Performance Test on a Me.109 F1/2 October 1941.

$ .... Computed by Meteor. H.Q. Levant.

FOURTH TEST OF ME. 109 G-2 (TROP).
PARTIAL CLIMBS.

1505-1615B HOURS. 17th JANUARY 1943.
ALTIMETER SET TO HEIGHT OF AERODROME 130 FT SHOWING 1014 MILLIBARS.

Indicated Altitude.	A.S.I.	Boost.	Revs. per Minute.	Ø Radiator Temperatures.	Altimeter Corrected for Calibratn.	$ Met. Computed Heights.	$ Met. Height.	A.S.I. Corrected for Calibratn.	X Position error Correction. 1.3	Corrected I.A.S.	Time Seconds.	Rate of Climb.
19000-21000	140 rptd.	falling 1.24	2700		18890 / 20860	19257 / 21403	2146	162	plus 6	148	39	3300
19000-21000	140	1.3	2700 ‡	90° C.	18890 / 20860	19257 / 21403	2146	142	plus 6	148	38	3390
19000-21000	180	1.3 falling.	2700		18890 / 20860	19257 / 21403	2146	162	plus 4	166	38	3390
19000-21000	180	1.3	2750		18890 / 20860	19257 / 21403	2146	181	plus 3	184	41	3140
19000-21000	200	1.3	2700	90° C.	18890 / 20860	19257 / 21403	2146	200	plus 2	202	46	2800

Ø .... Radiator flaps in "Automatic".

‡ .... Engine cutting due to low fuel pressure until electrical pump switched on.

$ .... Computed by Met. H.Q. Levant.

X .... Pitot position error taken from R.A.E. Report No. E.A. 39/11, Brief Performance Tests on Me.109.

FIFTH TEST OF ME. 109.G-2 (TROP).   1445-1530B HOURS. 19th JANUARY 1943.
ALTIMETER SET TO HEIGHT OF AERODROME 130 FEET SHOWING 1011 MILLIBARS.

Indicated Altitude.	A.S.I.	Boost.	Revs. per Minute.	Radiator Temperatures.	Altimeter Corrected for Calibratn.	Met. Computed Heights.	Met. Height Diffce.	Met. Computed Height.	A.S.I. Corrected for Calibratn.	Pitot Position Error.	Corrected I.A.S.	Time Seconds.	Rate of Climb.	True air speed.
26000–28000	150	1.0	2700		25860 / 27880	25700 / 27700	2000					55	2140	
27000					26870			26700						
26000–28000	150	1.0	2700	90° C.	25860 / 27880	25700 / 27700	2000					55.5	2120	
26000	250	1.17	2750	85° C.	25860			25700	250	plus 1	251			384
16500 +	246	1.3	2700		16400			16310	246	plus 1	247			319
16500 $	270	1.3	2700		16400			16310	270	plus 1	271			350

+ ... Radiator flaps wide open.

$ ... Radiator flaps shut.

SEVENTH AND EIGHTH TESTS OF ME.109.G-2 (TROP).   0935 - 1030B HOURS. 29th JANUARY 1943.

ALTIMETER SET TO HEIGHT OF AERODROME 130 FEET SHOWING 1015 MILLIBARS.

Indicated Altitude.	A.S.I.	Boost.	Revs. Per Minute.	Radiator Temper- atures.	Oil Pressure.	Altimeter Corrected for Calibratn.	Met. Computed Heights.	Met. Height Diffce.	A.S.I. Corrected for Calibratn.	Pitot Position Error.	Corrected I.A.S.	Corrected Time in Seconds.
130	170	1.3	2700	100° C.	O.K.				171	plus 3	174	$$$
2000	170				O.K.				171	plus 3	174	41
5000	170	1.3	2750	93° C.	O.K.	4980	5062		171	plus 3	174	94
10000	170	---	---	---	O.K.	9910	9911		171	plus 3	174	177
15000	170	1.3	2800	90° C.	O.K.	14890	14783		171	plus 3	174	276
20000	170	1.3	2800	100° C.	O.K.	19890	19777		171	plus 3	174	380
25000	170/150	---	---	---	O.K.	24880	24701		171/152			497
30000	150	0.88	2700	85° C.	O.K.	29930	29478		152	plus 4	156	688
33000	150	0.78	2700	95° C.	O.K.	---	---		152	plus 4	156	848
34000	150	0.74	---	---	O.K.	---	---		152	plus 4	156	953
35000	150	0.74	2750	---	O.K.	35190	34334		152	plus 4	156	1057

$$$ ... Airborne over aerodrome watch started.

ME. 109. G-2 (TROP).

Test.	Date.	Take off Run.	Wind.	Landing Run.	Wind.	Petrol.	Duration.
First	29th Dec. 1942					63 Gallons.	45 Minutes.
Second	30th Dec. 1942.					65 "	50 "
Third.	31st Dec. 1942.					—	50 "
Fourth.	17th Jan. 1943.	Approx. 270 yds.	12 m.p.h. 47 Deg. to take off Direction.	Approx.430 yds.	12 m.p.h. 20 deg. to landing Direction.	70 "	65 "
Fifth.	19th Jan. 1943	Approx.196 yds.	15 m.p.h. dead ahead.	Approx.440 yds.	15 m.p.h. dead ahead.	60 "	55 "
Sixth.	19th Jan. 1943.					15	15 "
Extra.	28th Jan. 1943.	250 yds.	23 m.p.h. Gusty.	270 yards.Swing.	23 m.p.h.Gusty.	27	20 "
Seventh.	29th Jan. 1943.	200 yard.	28 m.p.h. Ahead.	330 yards.	28 m.p.h.Ahead.	40 ?	50 "
Eighth.	29th Jan. 1943.	186 yards.	28 m.p.h. Ahead.	440 yards.	28 m.p.h.Ahead.	40	35 "

ME. 109 G-2 (TROP)

CALIBRATION OF ALTIMETER

STANDARD	First Test 29 Dec. '42 Instrument reads.	2nd. and subsequent Tests 30 Dec.'42 onwards. Instrument reads.
1000	1020	980
2000	2040	1990
3000	3060	2995
4000	4075	4000
5000	5080	5020
6000	6100	6040
7000	7110	7050
8000	8130	8070
9000	9150	9080
10000	10150	10090
11000	11150	11100
12000	12150	12100
13000	13140	13110
14000	14120	14110
15000	15115	15110
16000	16110	16080
17000	17090	17110
18000	18080	18110
19000	19050	19110
20000	20045	20120
21000	21040	21140
22000	22050	22120
23000	23060	23120
24000	24035	24150
25000	25070	25120
26000	26080	26140
27000	27095	27130
28000	28090	28120
29000	29115	29110
30000	30120	30070
31000		31030
32000		32000
33000		32930
34000		33850
35000		34810

METEOROLOGICAL   TEMPERATURES.

FEET	29th. Dec. Degrees. F	30th. Dec. Degrees. C	31st. Dec. Degrees C.	17th. Jan. Degrees C.	19th. Jan. Degrees C.	1200 B hrs. 29th. Jan.43 Degrees. C.
--				18.5	17.	
120	62 F.					13.5
500				17.25	14.5	12.2
1000		16		15.25	13.25	9.7
2000		12.5	11.	12.5	10.	6.9
4000		8.0		7.75	5.5	1.1
5000	40 F.		5.			
6000		5.0		2.5	1.	− 4.0
8000		1.0		.25	− .75	− 7.1
10000	24 F.	---	− 4.	---	− 4.	− 10.1
12000		− 1.5		− 4.5	− 8.	− 12.7
14000		− 4.2		− 7.	− 12.	− 15.8
15000	12 F.	− 6.0	− 9.			
16000		− 8.0		− 11.25	− 17.	− 20.3
18000		− 13.00		− 16.25	− 22.	− 22.4
20000	− 6 F.	− 17.8	− 21.	− 20.75	− 25.5	− 26.6
22000					− 29.	− 30.2
24000					− 34.	− 34.8
25000		− 31.7 +	− 29.			
26000					− 36.7 +	
30000	− 36.7°C+					− 44 ± 1° +
27000					− 40.0 +	
28000					− 42.7 +	
35000						− 50 ± 2° +

− + ... Estimated by Met.

FIGURE 1.

ME 109 G   TRUE AIR SPEED

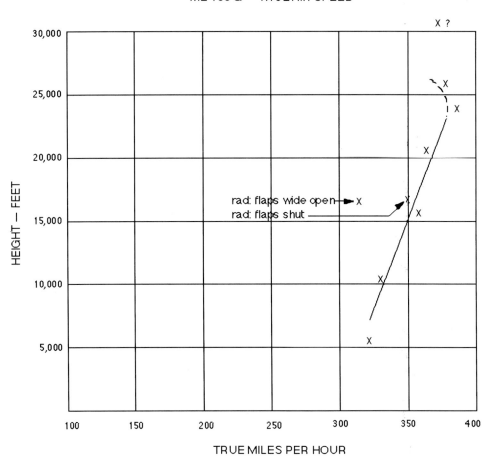

TRUE MILES PER HOUR

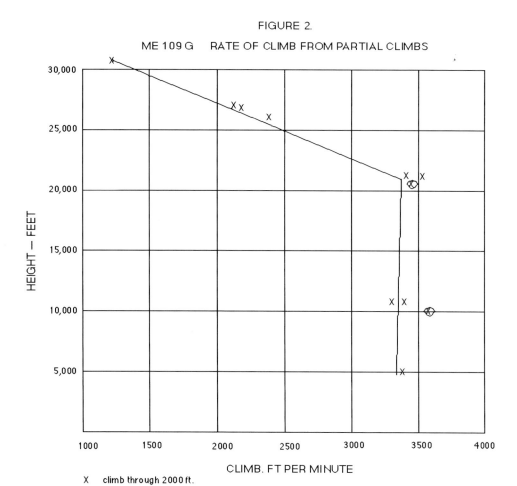

FIGURE 2.

ME 109 G   RATE OF CLIMB FROM PARTIAL CLIMBS

X   climb through 2000 ft.

⊗   from curves of 5 partial climbs

FIGURE 3.

ME 109 G-2 (TROP)

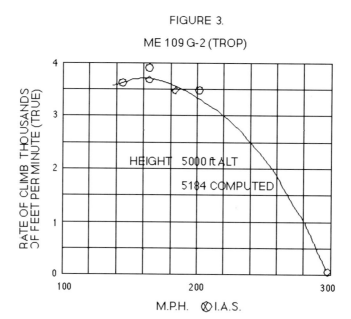

M.P.H. ⊗ I.A.S.

Met. state that rising air currents at 6000 ft up to 400 ft/min occurred on this day
⊗ I.A.S. not corrected for position error.

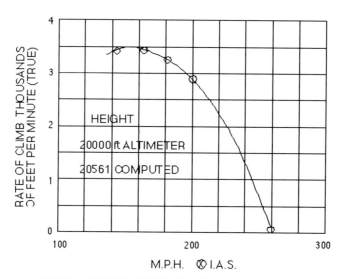

M.P.H. ⊗ I.A.S.

PARTIAL CLIMBS AT 5000 ft AND 20000 ft FULL THROTTLE
ENGINE BOOST 1.3 A.T.A.   R.P.M. 2700   WEIGHT   APPROX

JAN 17th 1943

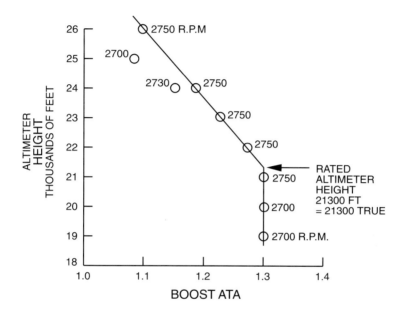

FIGURE 4.

THIRD TEST. DEC 31st 1942 CLIMB AT 200 M.P.H.
TO FIND RATED HEIGHT

FIGURE 5.

SEVENTH TEST   JAN 29th 1943

ME 109 G

TIME TO HEIGHT   FULL POWER CLIMB

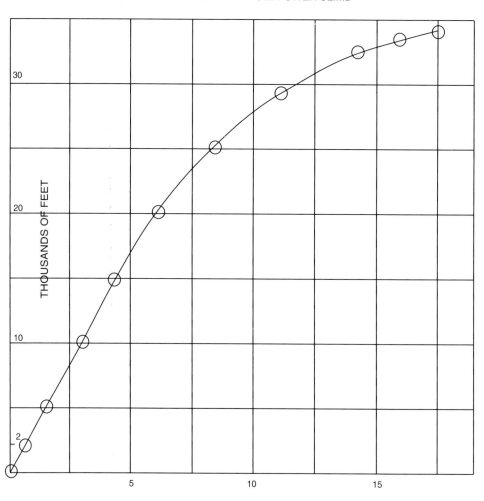

Appendix C

Maiden flight, Bf 109 G/G-USTV, 17 Mar 91

Introduction
1. The aircraft was in excellent condition having been completely rebuilt. There were no oil leaks or coolant leaks and the aircraft was clean externally and internally. A standard three-blade propeller and injector was fitted to the engine which was a Daimler-Benz DB 605.

Weather
2. The wind was 190/10 knots, temperature +15°C, QFE 996 mb, field elevation 203 feet, cloud 4/1900, 7/8ths at 2500 feet. Turbulence at low level was slight.

Cockpit
3. The cockpit had been faithfully renovated to reproduce the original. The panel below the glare shield held a modern radio. All the controls could be reached. However, it was necessary to understand German to read the labels. The circuit breaker panel was labelled separately.

Conditions Relevant
4. The aircraft was fuelled to full except for some 10 minutes of ground running, and the coolant and oil tanks were full. The C of G range was 2.776-3.0159 with the actual C of G at 2.8. The take-off weight was 2867.8 kgs. Rpm control was selected to Manual throughout the flight.

Take-off
5. The take-off proved difficult. Elevator trim was set at '0', rpm was in Manual with the rpm indicator set at 12 o'clock. The aircraft accelerated normally to about 40 knots using 1.15 boost and 2,400 rpm. At this point the wheels hit soft ground causing the aircraft to pitch forward and yaw hard to the left. Full back stick and full

right rudder only just controlled the aircraft. Although the pilot was not aware at the time, the propellor tips touched the ground. The grass strip felt very bumpy and the aircraft became airborne prematurely off a bump and was carefully flown away from the ground. The pilot was aware of holding an excessive amount of right rudder to keep the aircraft straight at this stage. This load eased as the aircraft accelerated and climbed away. The undercarriage retracted normally except the time to do so seemed excessive compared to other aircraft.

Climb and Cruise

6. The aircraft climbed briskly after take-off. Acceleration at 1,500 feet was also most noticeable. Cruise power of 2,100 rpm and boost of 0.9 ata gave a speed of 400 kph. Radiator temperature was steady at 80°C and oil temp. 80°C. Rpm were reduced by blipping the rpm switch on the throttle and it was easy to use. (Low switch equals low rpm.) Stick forces and movements to manoeuvre the aircraft were comfortable and easily applied. Aileron response was brisk and the aircraft was adequately damped in all axes. Quick aileron reverses could build up sizeable sideslip rapidly, but, on centralizing, yawing moment was damped in about 2 Hz with some feedback into the ailerons. Some right rudder was required throughout to hold the aircraft in trim, but PEs with rudder inputs were negligible. The noise level was comfortable, no fumes or smoke were apparent in the cockpit, the radio was easily heard and the engine was very smooth with a noticeably deeper beat than a Merlin.

Slow flight

7. The weather precluded full stalls, so minimum control speeds were flown in the clean and landing configuration with rpm control at 12 o'clock. The slowest speed achieved was 175 kph with gear down and full flap. The speed stability was excellent; the stick was well short of the aft stop; aft stick inputs produced an immediate pitch response and the ailerons remained effective. This speed, therefore, was used for landing.

8. At this stage difficulty was experienced with the undercarriage which appeared not to retract. After 'up' and 'down' selections without response, the undercarriage suddenly retracted after an up selection, but then could not be extended. At this time, electrical power to the engine instruments, the compass and undercarriage lights was also lost. The undercarriage could be seen to be extending very slowly by reference to the mechanical indicators, however, and by selecting a popped circuit-breaker, electrical power was recovered. The undercarriage eventually extended giving two green lights. The aircraft was recovered to the airfield holding a landing configuration, although the compass, which was at best difficult to use, became impossible having lost power.

The Landing

9. The landing was made after one low overshoot to check the approach. Adequate power was available to overshoot from a baulked landing at a late stage. During the landing flare, the aircraft exhibited the rather more prolonged float characteristics of a rounded wingtip Bf 109. Even allowing for this, the float seemed a little excessive and there may be some scope for reducing the threshold speed slightly from the 175 kph used, although this should be done with caution. It is suspected that the large metal propeller on this aircraft may be delivering some thrust even at idle and the pilot should be ready for a longer landing in this aircraft than other ME 109s.

Conclusions

10. The aircraft pitched hard forward during the take-off as the wheels hit soft ground. The left wheel apparently sank deeper than the right inducing a left yaw on the aircraft, however the yaw would have been exacerbated by:

    (a) The high rate of nose down pitch and the resulting precession from the large metal propeller.

    (b) The crosswind (although within limits) from the left.

    (c) The set rudder trim which needed more right bias.

11. In order to have better control over the nose down pitch, one division nose up trim is recommended rather than '0' used for this take-off.

12. The aircraft is in need of a small standby compass while it is engaged on flying operations.

13. Pilots should expect to use increased landing distance in this aircraft compared with other ME 109s.

Group Captain R. Hallam

Appendix D

Those who helped

Aeroquip Ltd/John Holder, Terry Cresswell
Anders Norling
Arnos Tap & Die Co Ltd
Avica Equipment/Meggit Aerospace/Mark Bolton, Peter Capstick
Barry Controls Ltd
Becker Flugfunkwerk GmbH
Berger Paints
Bestobell Aviation/Terry Wedlock
Bore Steamship Co Ltd
Robert Bosch GmbH
Wing Commander Paul Brindley
British Aerospace Civil Aircraft Ltd, Chester/Arnold Law
British Aerospace, Hatfield/Ken Wheeler
British Aerospace Military Aircraft Division, Hamble/Peter Knight
British Aerospace Military Aircraft Division, Warton/Andy Stewart,
    Tom Huckstable, Dennis Wade, Bernard Ward, Arthur Talbot.
British Airways, Heathrow
British Steel Corporation Special Steels
CASA/Señor Cervera
Rick Chapman
Chelton (Electrostatics) Ltd
Civil Aviation Authority/A. Jones, A. Bevin
Peter Cohausz
Heinz Dächsel GmbH
John Danes
De Soto Titanine POLC/Ian Wheelan, Brian Varley
Dowty-Rotol Ltd/F.J. Oliver
Dowty Seals Ltd/A.L. Harper
Drägerwerk Aktiengesellschaft
Dufay Titanine plc

Dunlop Ltd (Aircraft Tyres Division)/Charles Groves
Dunlop Polymer Engineering Division
Eagle Transfers
Façon GmbH
Finnair Cargo
Finnish Air Force
Flight Refuelling Ltd/Colin Thomas, Jack Green
FPT Industries
Freudenberg Angus LP/Barry Colledge
Squadron Leader G. Gatenby
GEC Aerospace, Fareham/Peter Taylor
GEC Avionics Ltd, Rochester/A.C. Smith
GEC Ferranti Defence Systems Ltd/John Dodds
Goodyear Tyre, Heathrow/P.N. Clark
Dr W. Gorniak
Jean Michel Goyat
High Temperature Engineers Ltd/Colin Roby, Ray Chamberlain
Hoffmann GmbH/Peter Ihrenberger
G.W. Howard
Humbrol Ltd
ICI Plastics Division
International Lamps
Irving Great Britain Ltd
Keski-Suomen Ilmailmuseo/P. Virtanen, K. Rasanen
Heinz Langer
Günter Leonhardt
Linread Aircraft Products/Ray McPhie
Lucas Aerospace/Peter Sharpe
Luftwaffenmuseum/P.J. Wiesner
Richard P. Lutz
Mann & Son (London) Ltd/Philip Mann
Mannesmann Hartmann & Braun AG
Henry Mapple & Sons
H. Marahrens
Messerschmitt-Bölkow-Blohm/Oskar Friedrich, Fred Owers, Werner
Blasel, Hans-Jochen Ebert, Herr Essenfelder
R. Meyer
Museum der Schweizerischen Fliegertruppe
Negretti Aviation Ltd/Arthur Gradding
Mick Oakey
Pattonair International Ltd
PPG Industries (UK) Ltd
Rolls-Royce Ltd, Bristol
Royal Air Force Lyneham, Northolt, Benson, Swanton Morley/Central
    Servicing Establishment, Brize Norton/NDT Section, St Athan, Abingdon
Royal Air Force Museum/Jack Bruce, Ray Funnell, John Wadham
Al Rubin
Sarma (UK)
Mike Schoemann
Wolfgang Segeta
Serck Heat Transfer/N. Ryder
Shamban Europa (UK) Ltd/Helga Whitmore

Don Silver
Bob Sinclair
SKF (UK) Ltd, Luton
Smiths Industries Defence Systems Ltd/E.A. Hooper
Squadron Leader P. Sowden
S.S. White Industrial Ltd/Don Sinfield
Stellite Company
Süddeutche Kühlerfabrik Julius Fr. Behr GmbH
Suntex
Technical Paint Services Ltd
Bob Thompson
Malcolm Towse
Fritz Trenkle
Christiaan Vanhee
Graeme Weir
Dr-Ing Andreas Weise
Ken West
John & Neil Westwood
Dipl-Ing Elmar Wilczek
Winter Bordgeräte Feinmechanik
Len Woodgate

Index

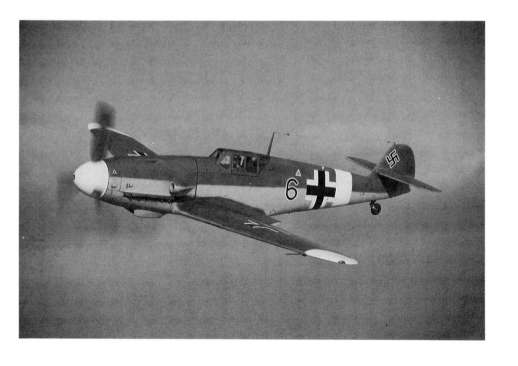